Advanced Methods of Machining

Advanced Methods of Machining

J. A. McGeough

Regius Professor of Engineering and
Head of the Department of Mechanical Engineering,
University of Edinburgh

London New York
Chapman and Hall

First published in 1988 by
Chapman and Hall Ltd
11 New Fetter Lane, London EC4P 4EE

Published in the USA by
Chapman and Hall
29 West 35th Street, New York NY 10001

Typeset in 10/12pt Times at
Keyset Composition, Colchester
Printed in Great Britain at the
University Press, Cambridge

ISBN 0 412 31970 5

British Library Cataloguing in Publication Data

McGeough, J. A. (Joseph Anthony), 1940–
Advanced methods of machining.
1. Materials. Machining
I. Title
670.42′3

ISBN 0 412 31970 5

Library of Congress Cataloging in Publication Data

McGeough, J. A.
Advanced methods of machining / J. A. McGeough.
p. cm.
Includes bibliographies and index.
ISBN 0 412 31970 5
1. Machining. I. Title.
TJ1185.M375 1988 88-9695
671.3′5—dc19 CIP

Contents

Preface

The mechanical cutting action that arises when a sharp tool makes contact with a workpiece of softer material has been the basis of established machine tool technology for almost two centuries. Indeed the principles on which these machines work continue to be the subject of many investigations.

In recent years, however, attention has turned towards the application of non-conventional methods of material removal. This book is devoted mainly to a presentation of such advanced methods. To that end an account is first presented of the early development of machine tools, and of some of the circumstances that have prompted the use of new methods for machining. Chapters 2 and 3 then deal with machining by, respectively, beams of electrons and gaseous ions. Another use of ions for material removal, by electrolytic dissolution, is the subject of Chapter 4. In Chapter 5 machining by light in the form of a laser beam is considered. In this case material removal effectively stems from heating of the workpiece. Further methods of machining that employ thermal effects are described in the next two chapters; Chapter 6 concerns the efficient machining which can be obtained utilizing the sparks generated by the breakdown of a liquid dielectric when a voltage is applied between two electrodes; and Chapter 7 deals with the machining obtained by applying a voltage between electrodes in a plasma of ionized gas.

Recent mechanical methods of machining are the subject of Chapters 8 and 9. In the former, material removal by the erosive effects of granular particles undergoing ultrasonic vibration is described. Chapter 9 deals with machining caused by a high-pressure water jet, enhanced mechanical action being achieved by the incorporation of abrasive particles. Chapter 10 is an account of other specialized methods of machining, some of which are well established whilst others have only recently been reported or have grown in significance.

The atomic and electronic effects on which most of these advanced methods are based form the substance of the Appendix, included for the benefit of those unfamiliar with this aspect of the subject. A unified notation for the symbols used throughout the book has proved to be a particularly difficult matter to resolve, as a consequence of the nature of the subject and of the variety of sources from which the information for the book has been drawn. In the end, the symbols used have been defined for each chapter, although the notation used throughout the book is largely uniform. On the other hand, and again despite the range of sources, all units have been expressed in the metric system, although the full SI system has been found to be difficult to implement. All chapters and the Appendix carry a bibliography and reference list for further consultation.

I am grateful to a great number of people for their assistance during the writing of this book. Mr J. McKee has been most assiduous with the literature search. For the high quality of the diagrams I am indebted to Mrs A. McQuillin. Mrs I. Duncan and Ms G. Moody helped with the typing of the manuscript through its various stages. Dr G. O. Goudie dealt cheerfully with its compilation, and with Dr J. W. Midgley read the entire manuscript, both making many useful suggestions for its improvement. Mr D. Clifton and Dr N. S. Mair have also commented on sections of the book in its draft form. A special expression of thanks is due to the many British industrial users of the processes described in this work. I have enjoyed a close working relationship with them over the years, and am much appreciative of their help in providing me with the benefit of their practical experience. The staff at Chapman and Hall, especially Ms M. Metcalfe and Mr M. Dunn, have liaised patiently with me throughout the preparation of the book.

Finally, my wife and family, Brenda, and Andrew, Elizabeth and Simon, have sustained their interest in this book from the outset and have thereby provided me with the encouragement needed to complete it.

J. A. McGeough
Edinburgh
August, 1987

The cover illustration of laser machining was adapted from a figure in a keynote paper presented by the late, respected Professor Raymond Snoeys and co-workers, at the 36th General Assembly of the International Institution for Production Engineering Research, 1986.

Notation

Symbol	Definition	Unit	Chapter
a	depth of penetration	mm	2
a_0	amplitude of tool motion	mm	8
A	process constant (electron beam machining)		2
A	atomic weight		4
A	area	mm^2	App*
A/zF	electrochemical equivalent		4
A^*	excited state		5
A_{21}	probability that a transition will occur per unit time, for spontaneous emission from levels 2 to 1		5
A_e	cross-sectional area of beam	mm^2	2
A_{max}	elongation amplitude	mm	8
b	process constant (EBM) (Equation 2.1)		2
B_{21}, B_{12}	proportionality constants for stimulated emission		5
B^*	excited state		5
c	specific heat of solid material	Jg^{-1} $degC^{-1}$	5
c	kerf width (Heat Affected Zone – HAZ)	mm	7
c	speed of sound; speed of light	cm/s	8
c_e	specific heat	Jg^{-1} $degC^{-1}$	4
C	energy needed to vaporize unit volume of the workpiece	J/mm^3	5
C	concentration of abrasive slurry		8
C_D	orifice coefficient		9
d	diameter of electron beam	mm	2
d	diameter of apertures in grid	mm	3
d	aperture diameter through which beam emerges	mm	5

*Appendix

Symbol	Definition	Unit	Chapter
d	diameter of embedded sphere	mm	8
d	mean diameter of grains	mm	8
d_{ac}	spacing between anode and cathode	mm	2
d_b	diameter of electron beam	mm	2
d_f	diameter of focused spot	mm	5
d_h	hydraulic mean diameter	mm	4
d_m	maximum diameter of abrasive grains	mm	8
$\bar{d}$	mean diameter of particles in the working gap	mm	8
D	diameter of mirror	mm	5
D	length of hole	mm	5
D	orifice diameter	mm	9
e	torch efficiency		7
e	electron charge	C	App
E	total energy to evaporate gold	J/g	5
E	energy output from lens	J	5
E	rate of energy transfer from plasma jet to workpiece	J/s	7
E	Young's modulus	MPa	8
E	electric field	V/m	App
E_e	energy of single electron beam pulse	J	2
E_i	energy level of excited atom, e.g. E_1, E_2, E_3	J	5
E_j	lower level of energy	J	5
E_m	magnetic energy	J	8
E_0	ground energy state at temperature T	J	5
E_w	mechanical energy	J	8
δE	decrease in energy	J	App
f	friction factor		4
f	electrode feed-rate (section 4.9)	m/s	4
f	focal length of lens	mm	5
f	frequency of changes in magnetic field (section 8.2)	Hz	8
f	frequency of tool vibration	Hz	8
f_p	pulse frequency	Hz	2
f_r	resonance frequency	Hz	8
F	Faraday's constant	96 500C	4

Symbol	Definition	Unit	Chapter
F	incident flux	W/cm^2	5
F	force applied	N	8
F_c	force of contact	N	8
F_s	static force	N	8
F_o	(constant) heat flux	W/cm^2	5
$F(d)$	statistical distribution		8
g_e	depth of material eroded per pulse; g is hole depth	mm	2
h	length of conductor (becomes inter-electrode gap width)	mm	4
h	Planck's constant (6.625×10^{-34})	Js	5 and App
h	impact indentation depth	mm	8
h_1	rate of energy provided by incident plasma jet	W	7
H	hardness		8
H	magnetic field intensity (Fig. 8.3)	A/m	8
H_o	maximum field intensity	A/m	8
H_t	hardness of tool		8
H_w	hardness of workpiece (brittle)		8
H_2	heat content of effluent	J/m^3	7
ΔH	energy to heat gold from ambient to an intermediate temperature between its melting point (1336 K) and 3000 K	J/g	5
HAZ	Heat Affected Zone (also kerf) width	mm	5, 7
I	current	A	2, 3, 4
I_e	space-charge limited emission current	A	2
I_o	intensity of beam at centre	W	5
$I(0)$	incident light intensity	W	5
$I(r)$	intensity of laser beam as a function of radius	W	
$I(x)$	light intensity at depth x	W	
J	current density	A/cm^2	App
J_e	maximum current density	A/cm^2	2
k	Boltzmann's constant, 1.38048×10^{-23}	JK^{-1}	2, 5
k	coefficient of thermal conductivity of solid material	$Wm^{-1} K^{-1}$	5
k	dielectric susceptibility		App

Symbol	Definition	Unit	Chapter
k_r	coefficient of magnetic coupling		8
kerf	*see* HAZ (width)	mm	5, 7
K	perveance	$AV^{-3/2}$	2
K	$2/\pi$ for a Gaussian beam		5
K	constant (accounts for spreading losses)		7
K	proportionality constant		8
l	external cylinder diameter, *see L*	mm	5
l	length of magnetostrictor core	mm	8
l	angular momentum quantum number		App
l_e	effective separation of grid	mm	3
Δl	increment in length l	mm	8
L	slot length	mm	2
L	internal cylinder diameter, *see l*	mm	5
m	mass to be removed by EBM, ECM, and LBM	mg	2, 4, 5
m	mass of ion	mg	3
mv	momentum of fundamental particles	Js/mm	App
m_e	average mass removed (EBM)	mg	2
m_l	magnetic quantum number		App
m_s	spin quantum number		App
n	atomic density	$atom/cm^3$	3
n	principal quantum number		App
n_e	number of electron beam pulses		2
n_{eL}	'linear' number		2
n_{ev}	volume number (EBM)		2
N	total number of particles		8
N	number of charged particles per unit volume	mm^{-3}	App
N	number of molecules per unit volume	mm^{-3}	App
N_o	population associated with energy level E_0	mm^{-3}	5
N_1, N_2, N_3	number of atoms per unit volume at levels E_1, E_2, E_3	mm^{-3}	5
p	pressure of cutting fluid	Pa	9
P	power density	W/cm^2	5
P	electrical power supplied to torch	W	7

Symbol	Definition	Unit	Chapter
P	polarization		App
q	charge on ion	C	3
q	ratio of hardness of workpiece to tool		8
Q	volumetric flow rate	m^3/s	9
r	radius of orbit	mm	App
r_a	recession rate of surface	m/s	4
r_f	radius of focused spot	mm	5
r_o	'Gaussian radius' (at which the intensity is $1/e^2$ of the central value of intensity)	mm	5
r_w	radius of contact indentation zone	mm	8
R	resistance	ohm	4
R	radius of beam or aperture	mm	5
R	resistance of conductor	ohm	App
R	surface roughness		6
Ra	surface roughness	μm	6, 7, 8
Re	Reynolds number		4
S	yield (Equation 3.2)	atom/ion	3
S	focused spot size	mm	5
S	skin depth, e.g. over which laser energy is absorbed	mm	5
t	time	s	5, 8
t	laser pulse length	s	5
t_i	time interval between pulses	s	2
t_m	time to drill hole in EBM	s	2
t_o	time for vaporization to start	s	5
t_p	time duration of electron beam pulse	s	2
T	temperature	K	5
T	material thickness	mm	10
T_m	internal temperature	K	5
T_o	initial temperature throughout	K	5
T_o	external temperature	K	5
T_1	temperature of incident plasma jet	K	7
T_2	effluent temperature	K	7
$T(x, t)$	temperature at depth x, time t	K	5

Symbol	Definition	Unit	Chapter
u_ν	energy density of radiation	J/m^3	5
U	electrolyte velocity	m/s	4
v	component velocity	m/s	5
v	velocity of water cutting jet	m/s	9
v	velocity of electron	m/s	App
v_m	mean velocity	m/s	App
v_c	velocity at which hole penetration is halved	m/s	5
$\dot{v}$	volumetric rate of metal removal	m^3/s	4
$\dot{v}$	material removal rate	m^3/s	8
$\overline{v}$	mean electrolyte velocity	m/s	4
V	potential difference between grids (Equation 3.1)	V	3
V	applied voltage	V	4
V	plasma flow rate	m^3/s	7
V	accelerating voltage	V	App
V_a	anode voltage	V	2
V_c	voltage (Figs 6.1 and 6.2)	V	6
V_d	volumetric rate of removal	m^3/s	8
V_o	accelerating voltage	V	2
V_o	volume fractured per grit	mm^3/min	8
$V(\theta)$	etch rate (Equation 3.2)	$atom/min^2$ per mA/cm^2	3
w	power density	W/cm^2	2
W_1	incident beam power	W	5
x	depth (see $I(x)$)	mm	5
x	distance of separation	mm	8
y	vertical co-ordinate	mm	8
y_o	amplitude of vibration of tool	mm	8
y_s	distance moved downwards by tool from mean position to position S	mm	8
z, Z	valency		4
z	error function variable		5
Ze	charge on nucleus	C	App

Greek symbols

Symbol	Definition	Unit	Chapter
α	absorption coefficient	m^{-1} or e.g. $10^5\ cm^{-1}$	5

Symbol	Definition	Unit	Chapter
δ_t	indentation in tool	mm	8
δ_w	indentation in workpiece		8
ϵ_m	coefficient of magnetostrictive elongation		8
ϵ_{ms}	saturated state of coefficient of magnetostrictive elongation		8
ϵ_o	dielectric permittivity	F/m	3
ϵ_o	permittivity of free space, 8.854×10^{-12}	F/m	App
θ	angular position	rad	8
$\theta_{\min}$	minimum beam divergence	rad	5
θ_s	angular position at S	rad	8
θ_t	lower limit of beam divergence	rad	5
$\delta\theta$	divergence of light beam	rad	5
κ	thermal diffusivity	cm^2/s	5
κ	thermal conductivity of metal	Wm^{-1} K	App
κ	specific conductivity (κ_e, of electrolyte)	$ohm^{-1}\ cm^{-1}$	5
λ	wave length	m	5
λ	wave length of radiation	m	App
μ	absolute viscosity	cP	4
ν	frequency of radiation	1/s	App
ν_{ij}	frequency of emitted radiation	1/s	5
ξ	error function variable		5
ρ	density of workpiece material	kg/m^3	2
ρ	resistivity	ohm cm	4
ρ	density of solid material	kg/m^3	5
ρ	density of magnetostrictor material	g/cm^3	8
ρ	density of cutting fluid	kg/l	9
ρ_a	anode metal density	kg/m^3	4
ρ_e	electrolyte density	kg/m^3	4
τ_x	brittleness criterion		8
ϕ	process constant (Equation 2.1)		2
ϕ	potential	V	4
ϕ	diameter	mm	5
ϕ	minimum aperture size	mm	10

1 Early progress in machining

From the earliest of times methods for cutting materials have had to be devised. Initially, materials had to be cut by hand, with tools made from bone, stick or stone. The oldest of these tools dates from over two and a half million years ago; these implements were used mainly to cut materials for clothing, cooking utensils, shelter and weapons. Later, elementary metals such as bronze and iron were employed in a series of advances in hand-tools over a period of almost one million years. Indeed, up to the 17th century, tools continued to be either hand-operated or mechanically driven by very elementary methods, typified by the simple tree-lathe shown in Fig. 1.1. By such methods, waggons, ships and furniture were manufactured as well as the basic utensils needed for everyday use.

The realization that water, steam and, later, electricity were useful sources of energy led to the production of power-driven machine-tools which rapidly replaced manual operations for many applications. Based on these advances and together with the metallurgical development of alloy steels as tool materials, a new machine-tool industry began to arise in the 18th and 19th centuries.

A major original contribution to this new industry came from John Wilkinson in 1774. He constructed a precision machine for boring engine cylinders, thereby overcoming a problem met with in the first machine tools, which were powered from steam. Initially steam engines had their cylinders bored out by machines originally designed to bore cannons. These machines were powered from water-wheels and were insufficiently accurate to meet the needs of the first machine-tools. Twenty-three years later, Henry Maudslay made a further advancement in machining when he devised a screw-cutting engine lathe. This machine incorporated a lead-screw for driving the carriage, geared to the spindle of the lathe; the use of the lead-screw enabled the tool to advance at a constant rate of speed so that accurate screw threads could be produced.

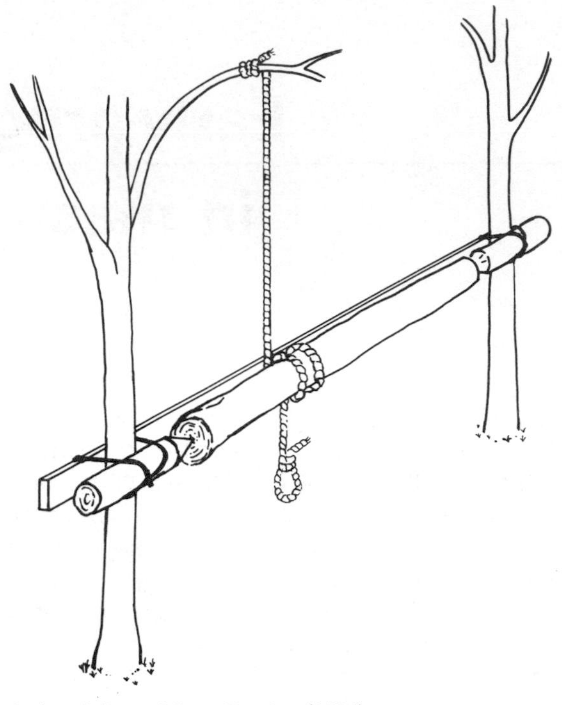

Fig. 1.1 Tree-lathe. Adapted from Perrigo (1916).

The lathe and boring mill are typical examples of turning machines, which are regarded as one of the seven main types of machine tool. As shown in Figs 1.2 and 1.3 the engine-lathe uses a single-point cutting tool for turning and boring. Excess metal is removed in the form of chips, for instance from the external diameter of a workpiece. In boring operations, a hole is enlarged or finished that has been previously cored or drilled.

James Nasmyth invented the second basic machine tool: a tool for shaping or planing (Fig. 1.2(b)). With this device a component, or workpiece, was clamped to a table and worked on by a cutter that had a reciprocating motion, enabling small surfaces to be planed, and keyways to be cut. Modern machine tools based on this principle are used to machine flat surfaces, grooves, shoulders, T-slots and angular surfaces with single-point tools. Nasmyth is also remembered for inventing, in 1839, the steam hammer for forging heavy pieces. He was a co-worker of Joseph Whitworth who introduced a further range of new or advanced machine tools. At the International Exhibition in 1862 the products from Whitworth's company occupied a quarter of the total space allocated to machine tools.

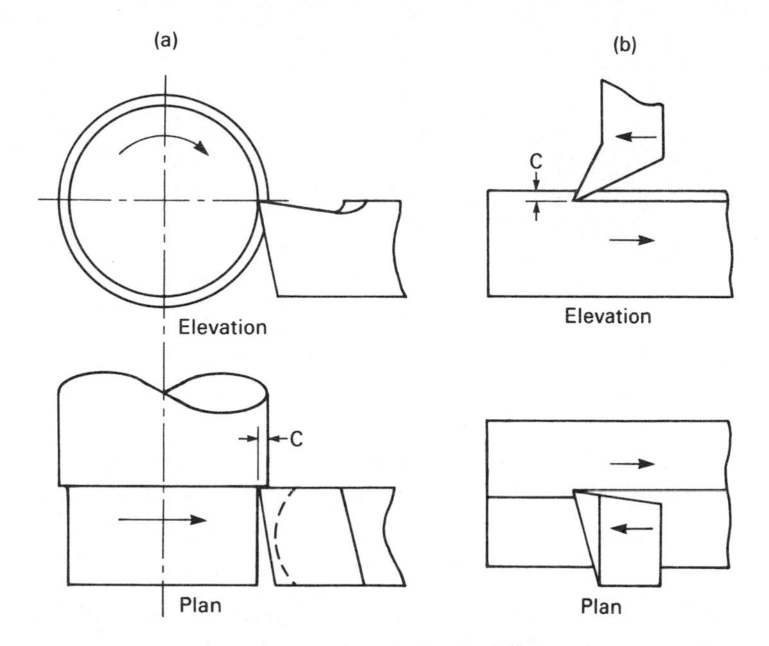

Fig. 1.2 Arrows denote direction of movement (a) Single-point lathe tool. (b) Tools for planing and shaping. (Depth of cut is indicated by *C*.)

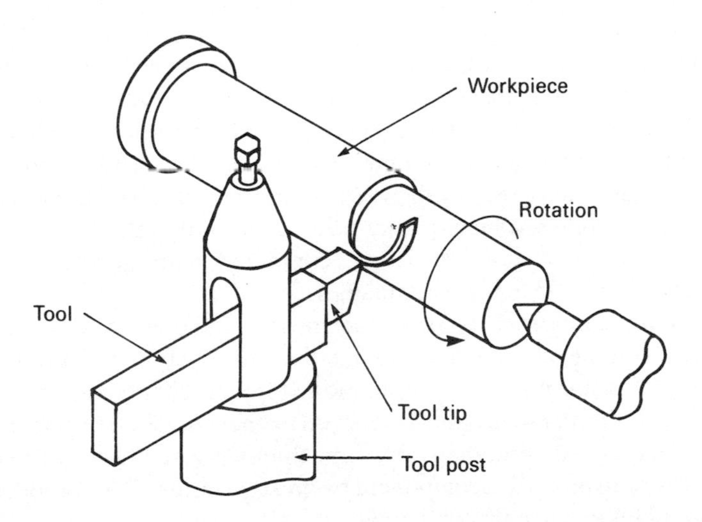

Fig. 1.3 Turning operation.

Fig. 1.4 Twist-drill.

The familiar drilling-machine is the third category of machine tools. As shown in Fig. 1.4 the device cuts holes with a twist-drill.

Another major advance came with the first milling machine, attributed to Whitney in about 1818, who used it in the manufacture of firearms. By feeding a workpiece against a rotating cutting tool termed a 'milling cutter', metal could be cut to a desired shape, such as grooves, dovetails and T-slots, as well as flat surfaces (Fig. 1.5). The first universal milling machine, constructed in 1862 by J. R. Brown, was employed to cut helical flutes in twist-drills.

Introduced in the late 19th century, a fifth type of machine-tool shown in Fig. 1.6 is the grinding machine: a rotating abrasive wheel or belt is used to remove small chips from metal parts which are brought into contact with it. An advanced form of this is 'lapping', where abrasive pastes or compounds are impregnated in a soft cloth which is rubbed against the surface of a workpiece to produce a high-quality surface finish or a finish to a very tight tolerance, as small as 0.00005 mm (compared with grinding in which accuracies of ± 0.0025 mm are attainable).

The two remaining main types of machine tools are power saws and presses. With the former, band-saws and circular disc-saws are used for cutting shapes in metal plates, for making internal and external contours, and for angular cuts. The forming of metal parts is achieved by the latter kind of machine tools which are equipped with a movable arm that is pressed against an anvil or base, in order to produce a component by one of a series of techniques such as shearing, blanking, bending and drawing.

The many notable developments of these seven basic machine tools include

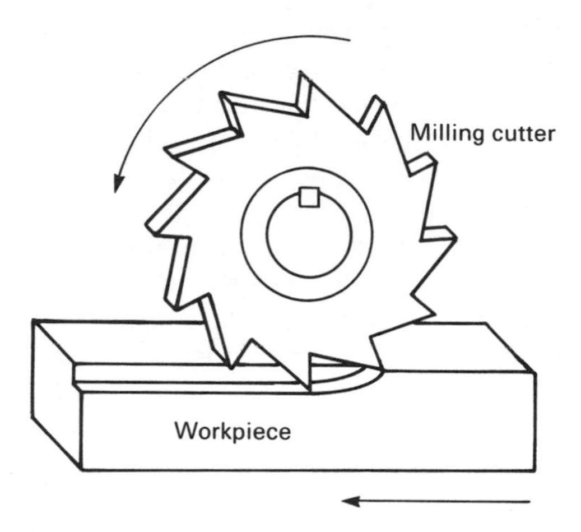

Fig. 1.5 Milling operation. Arrows denote direction of movement.

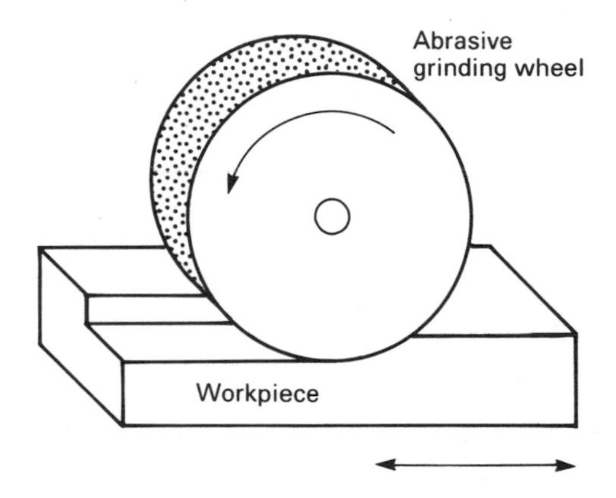

Fig. 1.6 Grinding. Arrows denote direction of movement.

the turret lathe made in the middle of the 19th century for the automatic production of screws. Another significant advance came in 1896, when F. W. Fellows built a machine that could produce almost any kind of gear. An example of the significance of early achievements in grinding technology came from C. N. Norton's work in reducing the time needed to grind a car crank shaft from five hours to 15 minutes. Multiple-station vertical lathes, gang drills, production millers and special purpose machines, for example, for broaching, honing and boring, are other noteworthy examples of advances in machine tool technology.

Such machines enable components to be cut or formed, usually to tolerances of ± 0.0025 mm (as noted above, accuracies of ± 0.00005 mm are possible with precision lapping machines). In summary, shaping by conventional machine tools relies on the following techniques: (*a*) cutting excess material from the part in the form of chips, (*b*) shearing the material, (*c*) squeezing parts to the desired shape.

In the latter part of the 19th and in the 20th century, machine tools became increasingly powered by electricity rather than steam. The basic machine tools underwent further refinement, for instance multiple point cutters for milling machines were introduced, and automated machine tool operations by computer control are now being rapidly developed. Even with these advances conventional machine tool practice still relies on the principle whereby the tool must be made of a material that is harder than the workpiece that is to be cut.

However, during the 19th and 20th centuries new physical phenomena have been uncovered which were just not known during the advent of these machine tools. In some cases the principles underlying some of these phenomena have been applied in the creation of entirely new methods of machining which do not rely on conventional metal removal for their operation. Applications for these new methods have then had to be sought. In others, these recently discovered physical principles have had to be invoked in order to overcome new industrial problems, such as those associated with the machining of new materials and alloys which are virtually intractable when tackled by established machine tools. In that respect Snoeys, Staelens and Dekeyser (1986) have drawn attention to the introduction of harder, tougher and stronger materials in manufacturing, including high-strength heat-resistant alloys, fibre-reinforced composites, ceramics, 'stellites' (cobalt-base alloys) and carbides. The mechanical properties of these materials often prove to be a formidable obstacle to their machining by conventional methods. Alternative methods of machining which are not defeated by the mechanical strength of materials are therefore of great interest. As will be seen in later chapters, the range of application of these new methods is often governed by other factors such as the electrical and thermal conductivity of the workpiece material, its melting temperature and its electrochemical equivalent.

In addition to the problems of machining very hard workpieces, modern materials often have to be cut to a complex shape, often in regions which are inaccessible by conventional methods. Advanced methods have been developed which can readily machine complicated shapes without encountering these difficulties.

Other limitations associated with mechanical methods became apparent in the 1960s during work on the machining of aluminium and other soft materials that were to be used as mirrors (Taniguchi, 1983). The maximum surface roughness tolerable was 0.01 μm, and machining had to be carried out to an accuracy of approximately 0.1 μm. Monocrystalline diamond cutting had to be used, and chips so produced were as small as or less than 1 μm. At these chip sizes the shear stress on the cutting tool becomes very large.

As Taniguchi points out, the distribution of movable dislocations in the metal crystals becomes almost zero, and the cutting forces then have to overcome the very strong bonding forces of the crystals. Very heavy tool wear can then be incurred.

The very fine finishes required also revealed the limitations of grinding, even with the finest of grits (such as carborundum). When single-point diamond cutting is used instead to produce the finish needed, the tool wear is again very great; and when glass, ceramics, and other hard brittle materials have to be machined in this way even more rapid wear of the diamond tools can be expected. Indeed Taniguchi explains how difficult it becomes to cut glass to a mirror finish by single-point diamond turning, and notes that this condition can normally only be obtained by special lapping and polishing techniques.

Even though highly sophisticated techniques have been devised by Bryan, and McKeown, and Sumiya, and their respective co-workers (Taniguchi, 1983), there remains a pressing need for new machining processes which might help to overcome the difficulties that arise.

The production of components for microelectronics has also led to a demand for methods for machining to extremely fine and precise limits. The major problems of precision machining of components to tolerances of 0.01 μm, with the dimensions of the parts as small as 1 μm, and surface roughness of 0.001 μm have been analysed by Taniguchi. In Fig. 1.7 he shows how the accuracies attainable have improved over the 20th century. A listing of the machines, processing and dimensional measuring equipment by which the resolution indicated can be obtained is included.

From the three general classifications of normal, precision and ultra-precision machining, the limitations of traditional methods become clear. Taniguchi concludes that these fine accuracies cannot be obtained by the mechanical removal of chips of material employed by conventional methods of machining. Instead, the precision required can only be achieved by advanced methods in which atoms or molecules of material are removed, either individually or in groups.

Many of these methods of machining are based on atomic and electronic principles which are discussed in the Appendix. In the light of the information presented there, we now proceed to examine the material removal that can be achieved with a beam of electrons.

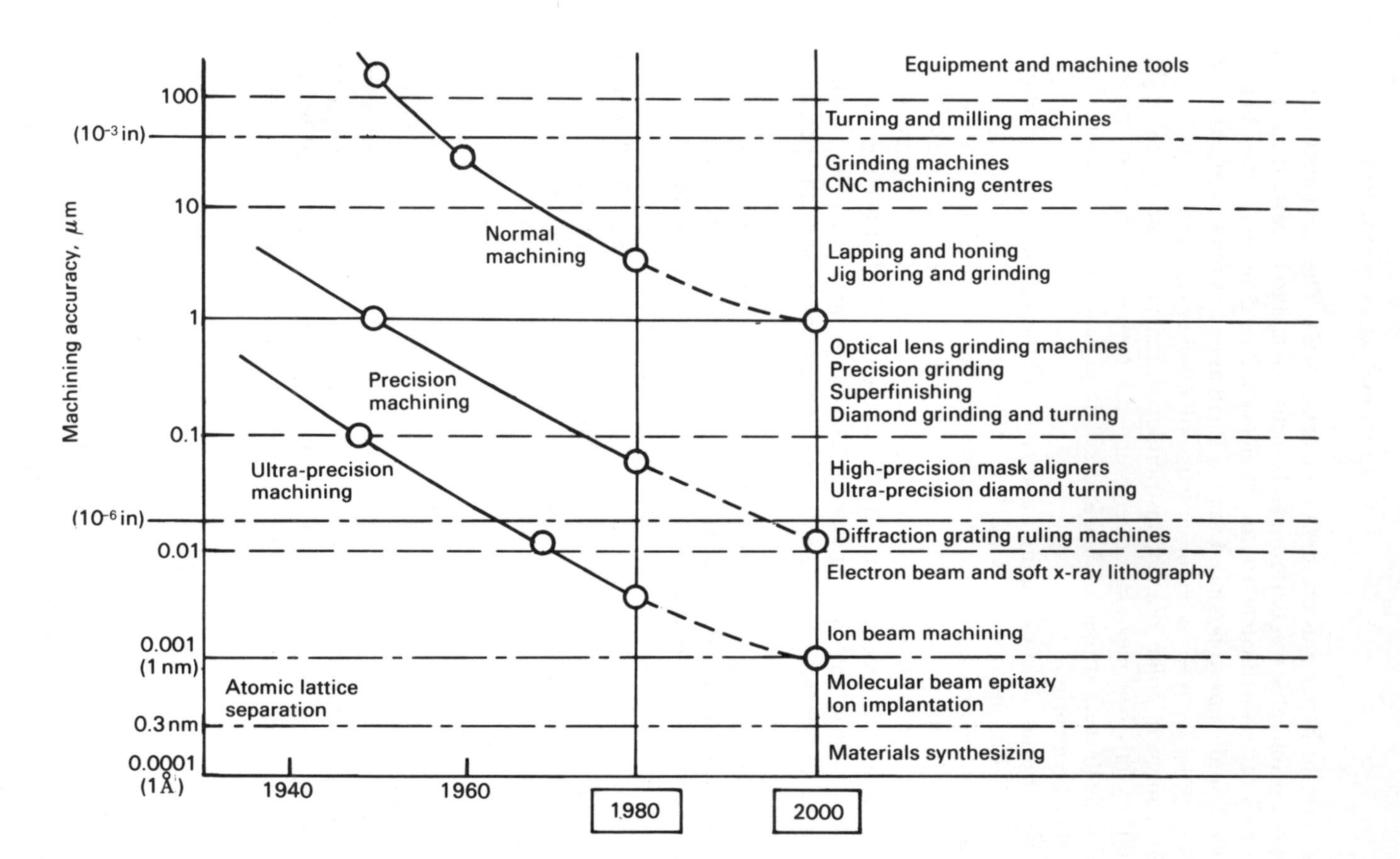

Fig. 1.7 Machining accuracies. (After Taniguchi, 1983 and McKeown, 1986.)

BIBLIOGRAPHY

McGeough, J. A., McCarthy, W. J. and Wilson, C. B. (1987) Electrical Methods of Machining in *Machine Tools*, Encyclopaedia Britannica, Vol. 28, pp. 712–36.
McKeown, P. A. (1986) High Precision Engineering and the British Economy (James Clayton Lecture, 1986), *Proc. I. Mech. Eng.*, **200** (76), 1–18.
Perrigo, O. E. (1916) *Lathe Design, Construction and Operation*, The Norman W. Henley Publishing Co., New York, p. 23, figure 1.
Rolt, L. T. C. (1986) *Tools for the Job*, Her Majesty's Stationery Office, London.
Snoeys, R., Staelens, F., Dekeyser, W. (1986) Current Trends in Non-conventional Material Removal Processes, *Ann. CIRP*, **35** (2), 467–80.
Taniguchi, N. (1983) Current Status in and Future Trends of Ultraprecision Machining and Ultrafine Materials Processing, *Ann. CIRP*, **32** (2), 1–8.

2 Electron beam machining (EBM)

2.1 INTRODUCTION

The final stages of the Appendix deal with the phenomena arising when electrons are generated within a vacuum chamber, similar to a cathode ray tube. Electron beam machining depends for its operation upon similar effects. Some of the earliest work on the utilization of the electron beam for material removal can be attributed to Steigerwald who designed a prototype machine in 1947. Modern electron beam machines work on the same principles.

2.2 BASIC EQUIPMENT

The main components of an EBM installation are shown in Fig. 2.1. They are housed in a vacuum chamber, evacuated to about 10^{-4} torr. The source of electrons is an 'electron gun', which is basically a triode consisting of a cathode, a grid cup negatively biased with respect to the cathode, and an anode at ground potential. The cathode is usually made of a tungsten filament, which is heated to between 2500 and 3000°C in order to emit the electrons. A measure of this effect is the emission current, the magnitude of which varies between 25 and 100 mA. Corresponding current densities lie between 5 and 15 $\mathrm{A cm^{-2}}$. This quantity however is determined by a range of factors, including the type of cathode material and its temperature (see below). The size of the emission current is also influenced by a high voltage, usually about 150 kV, which is applied between the cathode and anode in order to accelerate the electrons in the direction of the workpiece. Fig. 2.2 illustrates how the emission current increases with cathode temperature and accelerating voltage, when that electrode is made of a tungsten filament.

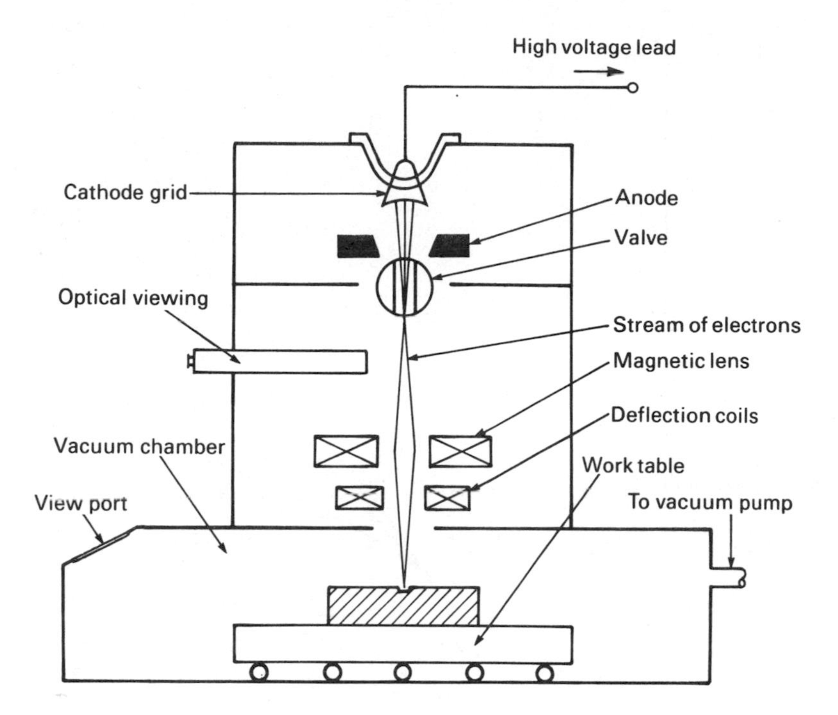

Fig. 2.1 Components of electron beam machine.

After acceleration, the electrons are focused by the field formed by the grid cup so that they travel through a hole in the anode. On its exit from the anode cavity, the electron beam is re-focused by a magnetic or electrostatic lens system. By this means, the beam has its direction towards the workpiece kept under control. The electrons maintain the velocity imparted by the accelerating voltage, until they strike the workpiece specimen, over a well-defined area, typically 0.025 mm in diameter. There the kinetic energy of the electrons is rapidly translated into heat, causing a correspondingly rapid increase in the temperature of the workpiece, to well above its boiling point. Material removal by evaporation then occurs. With power densities of the order of 1.55 MW mm^{-2} involved in EBM, virtually all engineering materials can be machined by this technique.

Figure 2.3 (a), (b) and (c) shows details of an industrial electron beam machine. This particular unit can be used for welding as well as machining. Accurate manipulation of the workpiece coupled with precise control of the beam is reported to yield a process that can be fully automated.

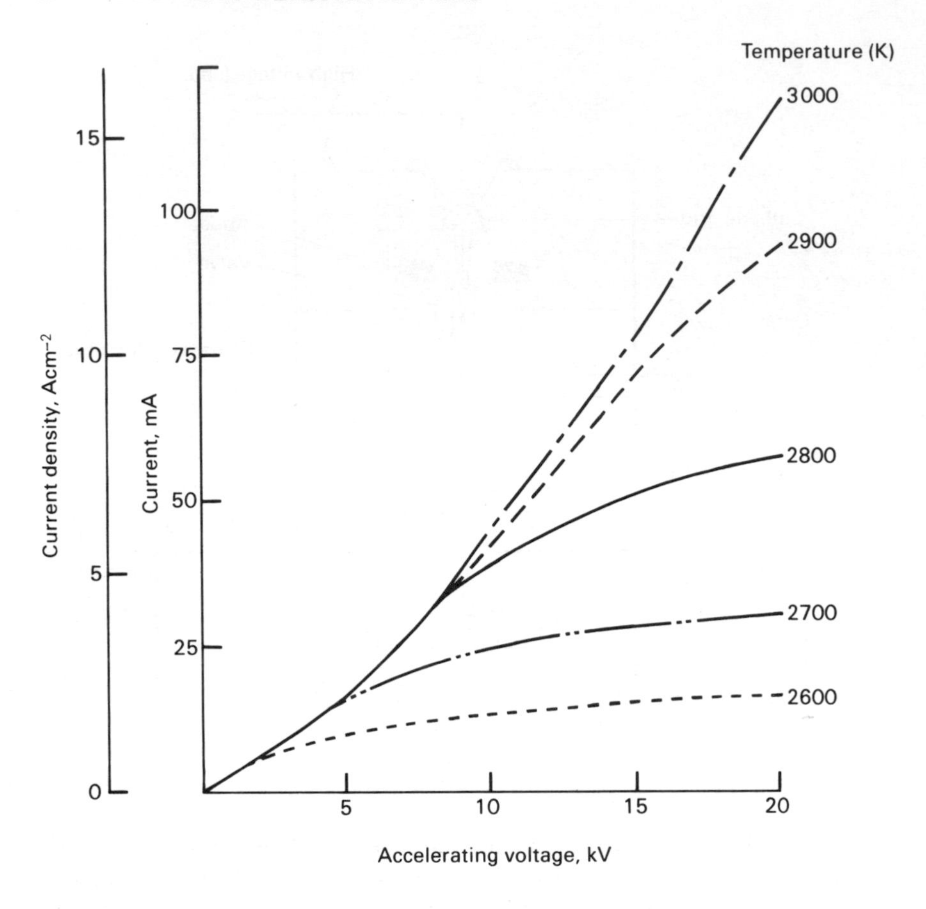

Fig. 2.2 Variation of emission current with cathode temperature and accelerating voltage. (Adapted from Kaczmarek, 1976.)

2.3 EMISSION CURRENT

Owing to the significance of the emission current in EBM, further attention to this process variable is useful. Several theoretical expressions for the corresponding emission current density are available which describe different aspects of the electron beam process. For example, when the voltage gradient in front of the emitter is sufficiently high to draw off the electrons, a condition known as 'temperature-limited emission', the maximum current density J_e can be expressed as

$$J_e = AT^2\exp(-b\phi/kT) \tag{2.1}$$

where A, b and ϕ are constants for the material, T is its temperature and k is Boltzmann's constant.

An alternative useful expression for the emission current density is

$$J_e = 2.33 \times 10^{-6} \frac{V_a^{3/2}}{d_{ac}^2} \tag{2.2}$$

This can be employed when the voltage gradient is less than that described above. In this case, the emission of the electrons is hindered by a negative space charge in the vicinity of the cathode. That is mutual electron repulsion takes place there. In equation (2.2) above, which is known as the Childs–Langmuir relation, V_a is the anode voltage, and d_{ac} is the spacing between the anode and cathode. This equation applies for plane, parallel electrode configurations. For other anode–cathode geometries, a general expression for this 'space-charge limited' emission current I_e is

$$I_e = KV_a^{3/2} \tag{2.3}$$

where K is called the 'perveance', a quantity largely dependent on the geometry. Clearly when high emission currents are desired, large values for K should be sought. Typically, this quantity is about $2 \times 10^{-6} AV^{-3/2}$. The current density J_e and current I_e are related through the expression

$$J_e = \frac{I_e}{A_e} \tag{2.4}$$

where A_e is the cross-sectional area of the beam.

In the region where the beam of electrons meet the workpiece, their energy is converted into heat. The way in which the focused beam penetrates the workpiece is still not completely understood, owing to the complexity of the mechanisms involved; however, it is known that the workpiece surface is melted by a combination of electron pressure and surface tension. The melted liquid is rapidly ejected and vaporized to effect material removal. The temperature of the workpiece specimen outside the region being machined is reduced by pulsing the electron beam. Pulse frequencies seldom exceed 10^4 Hz.

2.4 THEORETICAL CONSIDERATIONS

The energy E_p of a single pulse of duration t_p for an accelerating voltage of V_o is given by

$$E_p = V_o I_e t_p \tag{2.5}$$

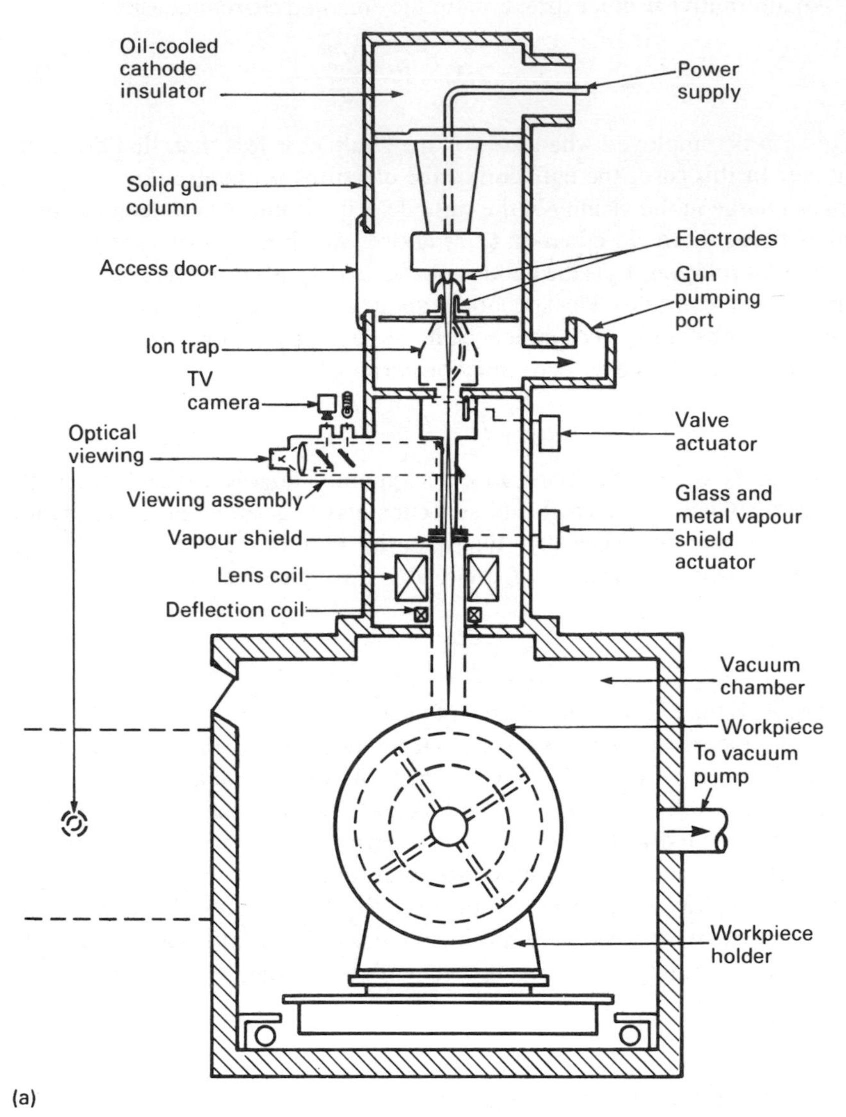

Fig. 2.3 Industrial electron beam machine. (Courtesy of Wentgate Dynaweld Ltd.)

(b)

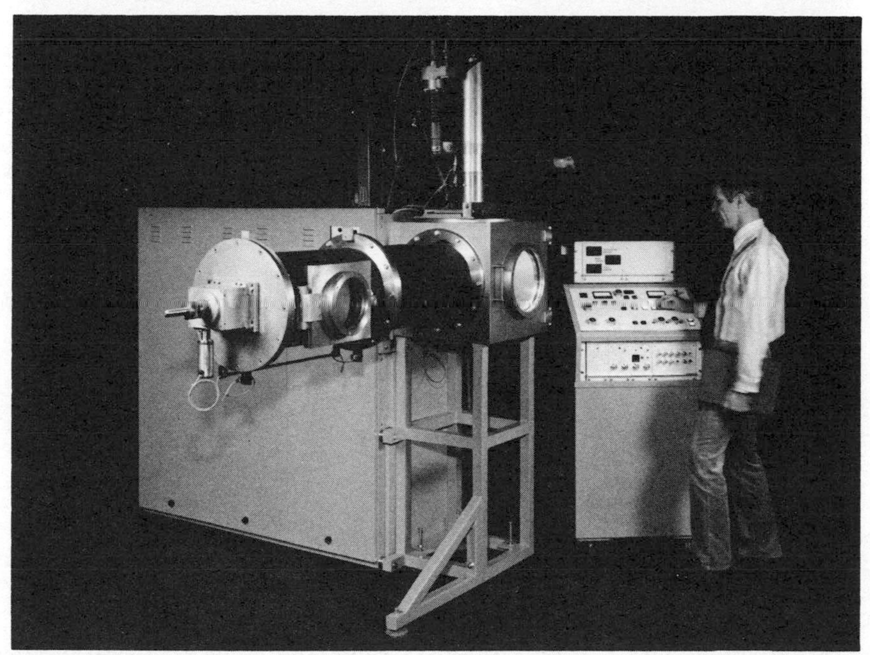

(c)

its frequency f_p being

$$f_p = \frac{1}{t_p + t_i} \tag{2.6}$$

where t_i is the interval between pulses.

The power density w

$$w = \frac{I_e V}{A_e} \tag{2.7}$$

has a significant effect on the rate of material removal, since, through the emission current, the temperature of the workpiece material and its rate of heating depend on it.

An early attraction of EBM was the comparatively large depth-to-width ratio of material penetrated by the beam with applications of very fine hole drilling becoming feasible. To that end, expressions for the depth to which the electron beam penetrates the material received close attention. Unfortunately the analytic expressions produced have often been complicated and conflicting. For instance, one group proposed that the depth a of penetration may be given by

$$a \propto w V^{-1/2} d^{3/2} \tag{2.8}$$

where d is the diameter of the electron beam, w is the power density and V is the accelerating voltage. Their proportionality constant was made up from complicated expressions involving various (mainly thermal) properties of the workpiece material. Kaczmarek (1976), on the other hand, quotes a simpler alternative expression, in which the depth penetrated varies thus

$$a \propto V^2 \tag{2.9}$$

although he does not define completely the proportionality constant.

Fig. 2.4 illustrates the type of hole formed in an alloy steel after a single pulse of EBM.

The depth of eroded material per pulse g_e may be related to the average mass m_e of material removed, by the expression

$$m_e = \frac{\pi d_b^2 g_e \rho}{4} \tag{2.10}$$

where ρ is the density of the workpiece material, and d_b is effectively the diameter of the electron beam at contact with the machined surface.

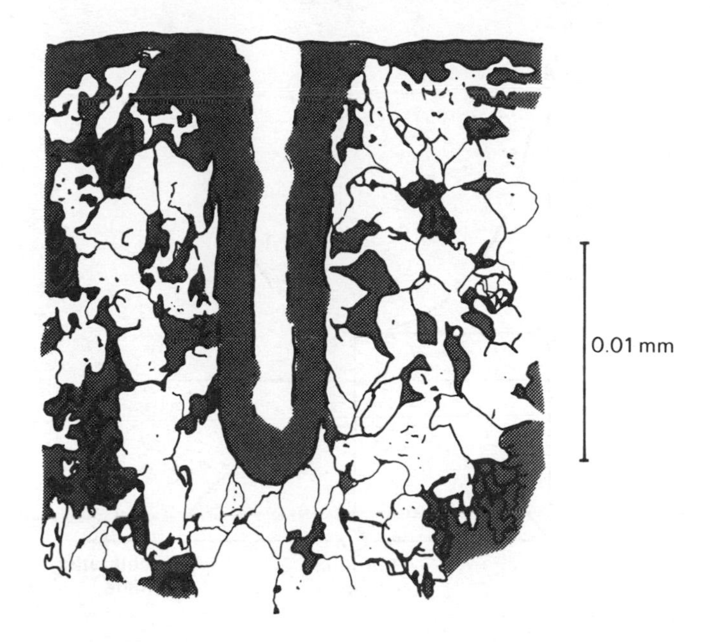

Fig. 2.4 Cross-section of a cavity in chromium-molybdenum steel formed by a single pulse. (Adapted from Kaczmarek, 1976.)

On the basis of these assumptions and relationships, Kaczmarek has derived an expression for the average depth g_e of material removed by a single pulse. The derivation is complicated, as indeed is the equation itself. Instead, therefore, of quoting this expression here, it is more relevant to the discussion if the main conclusions from his analyses are reported.

Kaczmarek's analysis reveals the existence of an optimum value of the accelerating voltage V_o for which the average depth g_e eroded by a single pulse reaches a maximum value. From his work the number of pulses n_e required to erode a hole of depth g can also be obtained, since

$$n_e = g/g_e \tag{2.11}$$

The graphical interpretation of his results shown in Fig. 2.5 indicates that there is a minimum number of pulses n_e associated with an optimum accelerating voltage. In practice the number of pulses to produce a given hole depth is usually found to decrease with increase in accelerating voltage.

His analysis also reveals that, for a fixed set of process conditions, the number of pulses required n_e increases hyperbolically, as the depth g increases. In practical terms this result means that when a certain depth has been reached, any further EBM to deepen the hole would require a very large increase in the number of pulses.

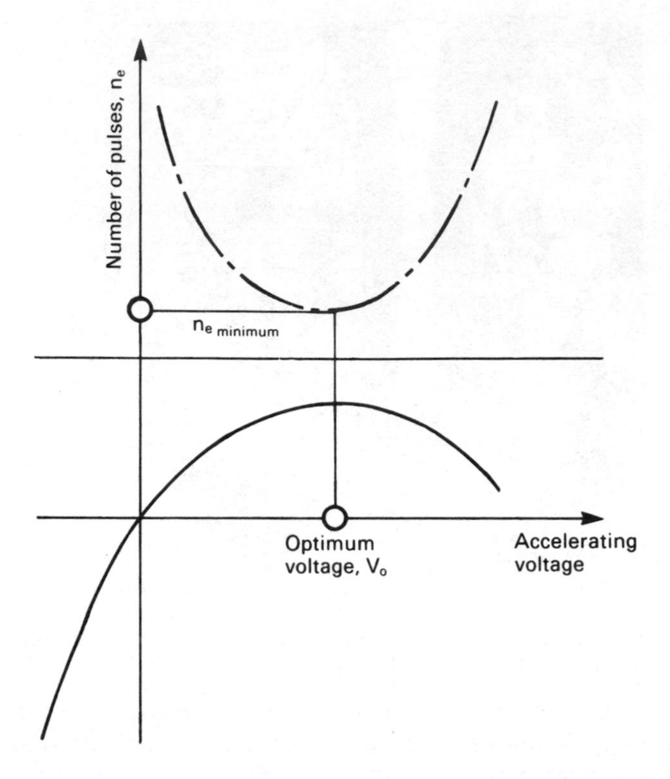

Fig. 2.5 Dependence of number of pulses required on accelerating voltage. (After Kaczmarek, 1976.)

The time of EBM, t_m, needed to drill a hole is given by

$$t_m = n_e/f \tag{2.12}$$

where f is the pulse frequency.

When a slot of depth g is to be formed by EBM, the machining time required in this case is found to be given by

$$t_m = \frac{n_e t_p L}{d_e} \tag{2.13}$$

where L is the slot length, and d_e is the diameter (width) of the electron beam at the region of incidence with the workpiece and t_p is the duration of the pulse.

2.5 RATES OF MATERIAL REMOVAL IN EBM

Electron beam machining rates are usually evaluated in terms of the number of pulses required to evaporate a particular amount of material. The use of electron counters registering the number of pulses enables ready adjustment of the machining time to produce a required depth of cut.

To that end, two types of pulse numbers are used. Firstly, the 'volume number' n_{ev} is used for evaluating slotting by EBM, that is, cutting-off or cutting-into materials.

$$n_{ev} = m/m_e \tag{2.13a}$$

where m and m_e are, respectively, the mass of material to be removed and the mass of material removed by a single pulse. Secondly, a linear number n_{eL} is adopted for hole sinking by EBM:

$$n_{eL} = g/g_e \tag{2.14}$$

where g is the depth of hole required and g_e is the depth of cavity sunk by a single pulse. Studies of the EBM of different metals have revealed that as might be expected their boiling point and thermal conductivity play a very significant role in determining how readily they can be machined. Other thermal properties, such as electrical conductivity of the material, are additional factors. The complexity of the relationship between the material properties and machinability renders difficult any fully quantitative analysis of removal rates in EBM.

Experiments still have to be performed to obtain representative values of machining rates. Table 2.1 shows typical results for tungsten and aluminium.

Table 2.1 Removal rates in EBM (power 1 kW) (Adapted from Bellows, 1976)

Material	*Volumetric removal rate* ($mm^3\ s^{-1}$)
Tungsten	1.5
Aluminium	3.9

Experimental results on the dependence of pulse numbers on accelerating voltage given in Fig. 2.6 show that increasing the hole depth requires a much greater rise in the number of pulses at low voltage, due mainly to a relative rise in heat losses resulting from conduction and melting of the adjacent metal layers. For a given number of pulses little improvement in material removal rate is obtained from increasing the accelerating voltage above 120 kV.

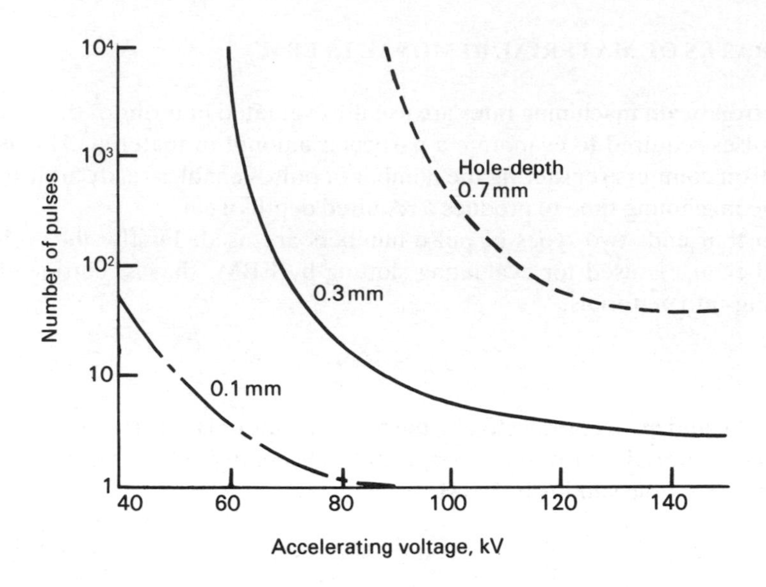

Fig. 2.6 Effect of accelerating voltage on number of pulses required. (After Kaczmarek, 1976.)

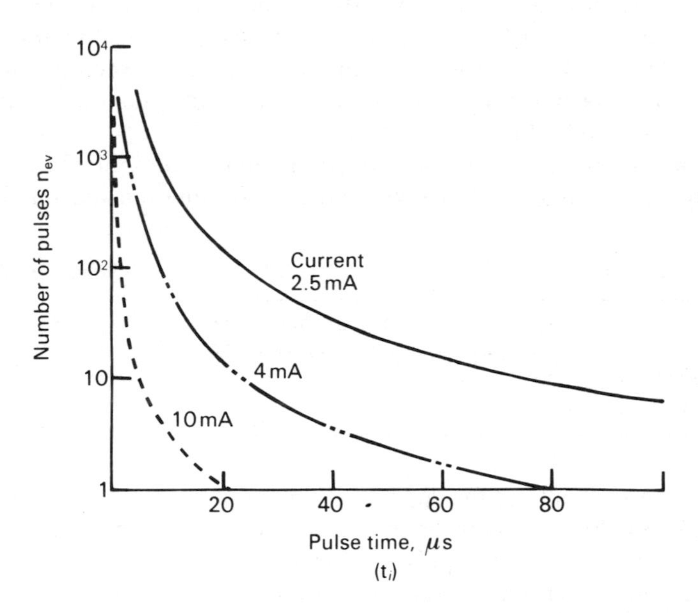

Fig. 2.7 Dependence of number of pulses needed on pulse duration. (After Kaczmarek, 1976.)

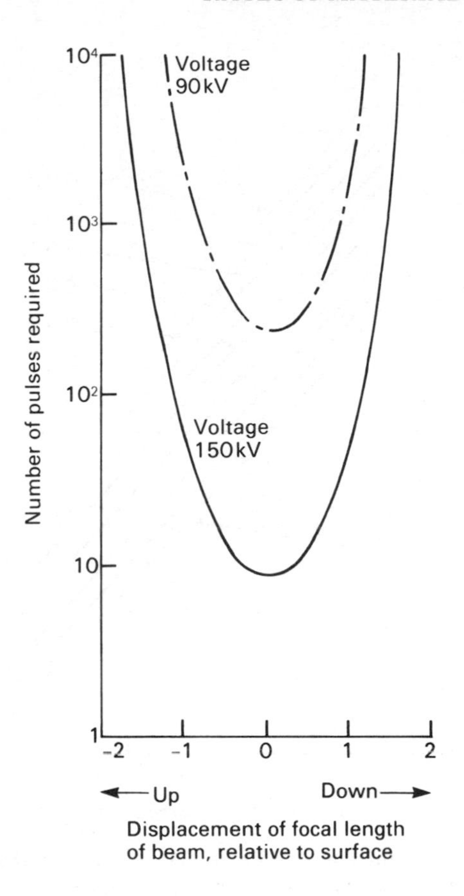

Fig. 2.8 Effect of displacement of focal length of beam relative to upper surface of workpiece on number of pulses. (Thickness of specimen is 2 mm.) (After Kaczmarek, 1976.)

Figure 2.7 indicates that an increase in the pulse duration, with a corresponding rise in the pulse energy made available, reduces the number of pulses needed to obtain the required machining result.

Following much evidence that the electron beam has to be focused very carefully if the best machining rates are to be obtained, Kaczmarek quotes reports of an *optimum working distance*, at which a minimum number of pulses is required, as illustrated in Fig. 2.8. A focal point just below the upper surface of a workpiece is sometimes found to be most effective.

Figure 2.9 shows how the drilling-rate by EBM (in holes per second) decreases with increase of the thickness of the workpiece and with increase of

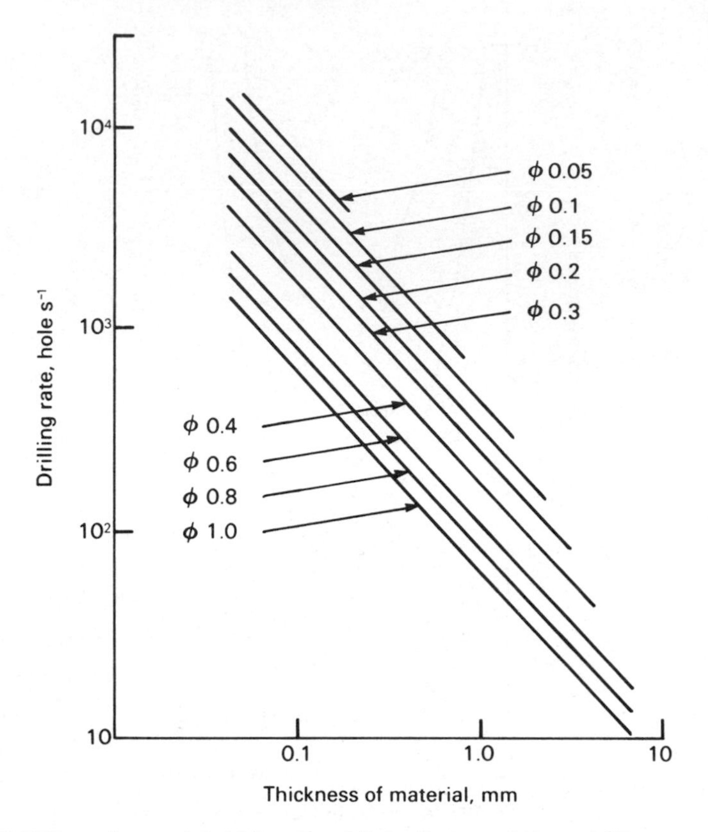

Fig. 2.9 Effect of material thickness and hole diameter (ϕ) on drilling rate. (Workpiece materials: steel and nickel alloy.) (After Bellows and Kohls, 1982.)
ϕ denotes hole diameter (mm)

the diameter of the hole to be produced. The relevance of these results to micromachining of thin materials will be discussed below.

2.6 SURFACE ROUGHNESS OF WORKPIECE IN EBM

The quality or surface roughness of the edges produced in EBM depends greatly on the type of material. Local pitting of the surface is a common occurrence, the extent of which is influenced by the thermal properties of the workpiece, and by the pulse energy or charge. Figure 2.10 illustrates how surface roughness can increase with pulse charge for a range of common materials, nickel, titanium, carbon, gold and tungsten.

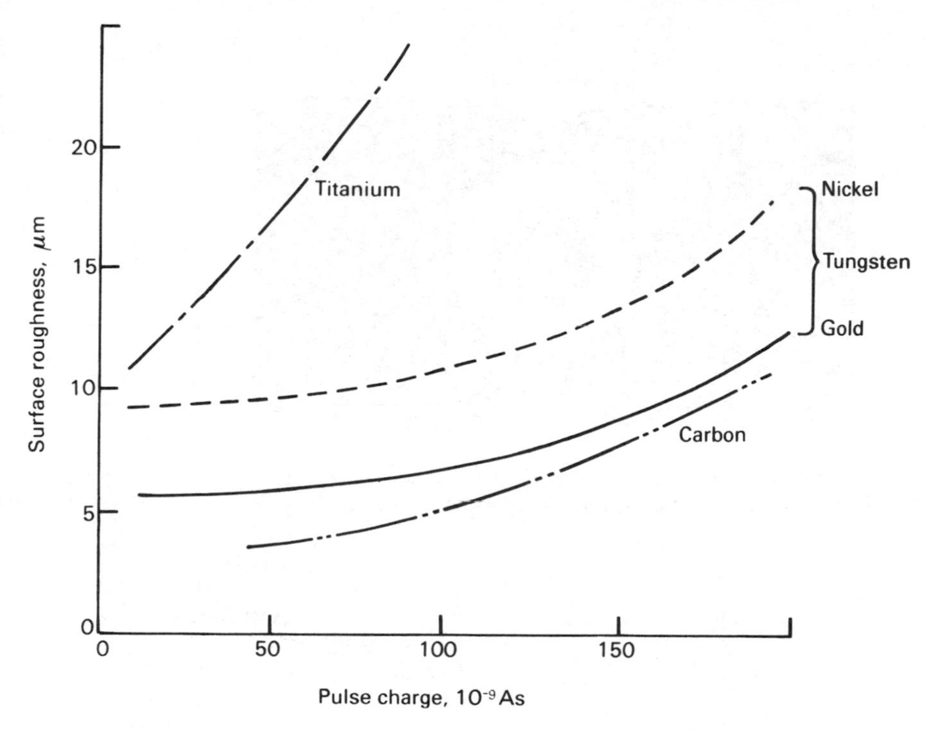

Fig. 2.10 Surface roughness as a function of pulse charge. (Adapted from Kaczmarek, 1976.)

2.7 HEAT-AFFECTED ZONE

As the evidence in Fig. 2.11 indicates, the surface layers of materials treated by EBM are affected by the high temperatures of the focused beam, illustrated by the white ring surrounding the hole. Further evidence is given in Fig. 2.12 which shows how the diameter of the damaged layer increases with pulse duration, as well as hole diameter. The heat-affected zone can be as much as 0.25 mm in EBM.

2.8 APPLICATIONS OF EBM

Applications lie in the following main areas (*a*) drilling, (*b*) perforating of sheet, (*c*) pattern generation associated with integrated circuit fabrication (with which milling is also associated).

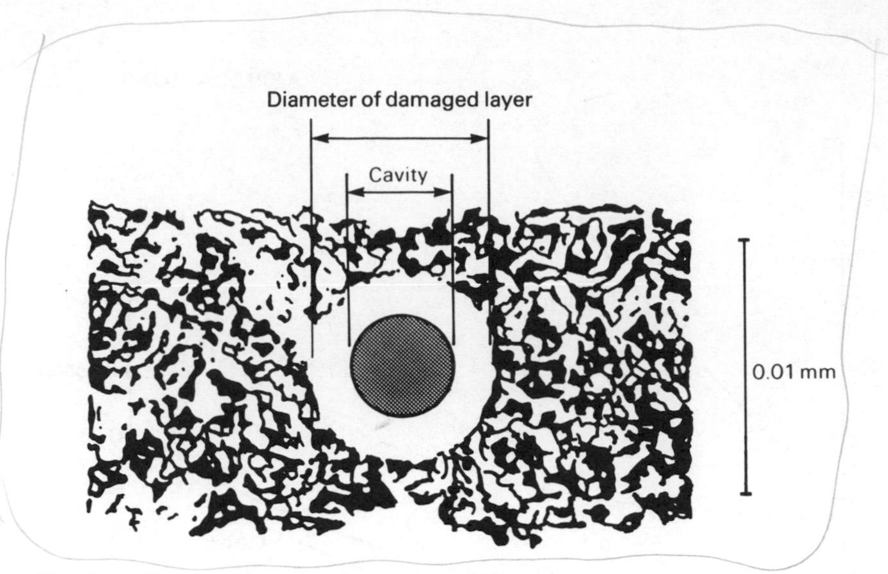

Fig. 2.11 Cross-section of cavity made by a single electron beam pulse in chromium-molybdenum steel. (After Kaczmarek, 1976.)

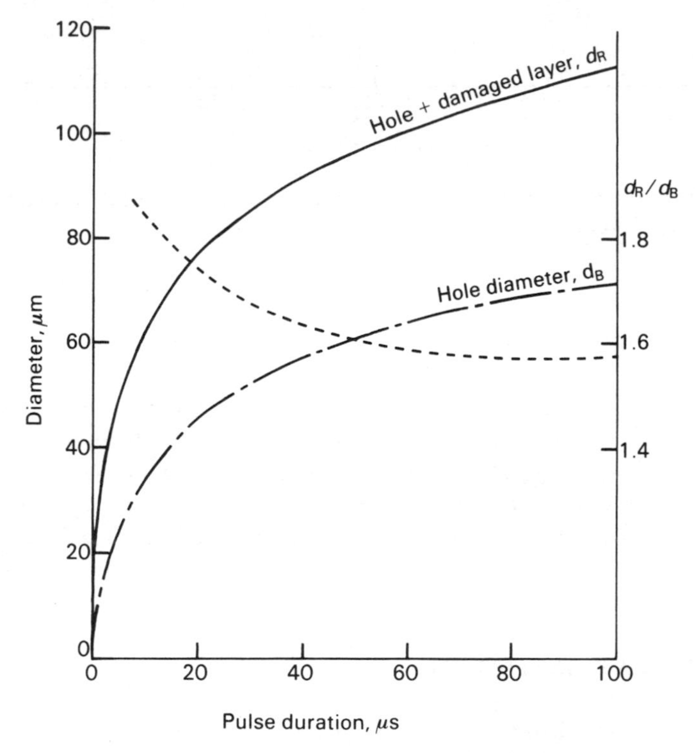

Fig. 2.12 Effect of pulse duration on width of hole drilled by EBM. (After Kaczmarek, 1976.)

2.8.1 Drilling

Steigerwald and Meyer (1967) gave early consideration to the use of EBM for hole-drilling. They concluded then that for successful applications, improved reproducibility, greater working speeds, and deeper holes of accurately controlled shapes were all needed. More recently Boehme (1983) discussed drilling applications with electron beam machines fitted with systems for numerically controlling the beam power, focus and pulse duration, and mechanical motion. As a result cylindrical and other configurations, such as conical – and barrel – shaped holes, of various diameters can now be drilled with consistent accuracy at rates of several thousand holes per second. The prospect for drilling inclined holes, at an angle of 15°, was also discussed.

Boehme claims that the largest diameters and depths of holes that can be accurately drilled by EBM are, respectively, 1.5 mm and 10 mm, and that the ratio of depth to diameter is normally in the range 1:1 and 1:15. For example, stainless steel plate 0.25 mm thick has been drilled with 0.2 mm holes (Birnie and Champney, 1967).

For deeper holes, in the range 2.5 to 7.5 mm, Steigerwald and Meyer (1967) emphasize the need for a stable power supply that can emit the required groups of pulses and for a well-controlled beam of closely defined diameter, the angle of aperture of which has a strong bearing on the shape of hole produced. Under laboratory conditions, holes of about 19 mm were achieved by their team.

Figure 2.13 (a and b) show samples of drilling obtained by them, for different materials. A further useful summary of the characteristics of EBM of various materials is given in Table 2.2. Fig. 2.14 shows the cross-section of holes drilled by EBM under the results (a) of Table 2.3.

2.8.2 Perforation of thin sheet

The sheet to be perforated is usually lined with an auxiliary material. The electron beam first penetrates through the sheet forming a vapour channel within the fused material, and then enters the auxiliary lining. An eruption of vapour occurs, causing ejection of molten material (Boehme, 1983). For perforation by EBM to be economically acceptable, 10^4 to 10^5 holes per second have to be produced. Thus single pulses lasting only a few μs are needed. In some applications the sheet or foil metal is stretched on a rotating drum, which is simultaneously shifted in the direction of its axis. Rows of perforations following a helical line are thereby produced. Manipulators capable of linear and rotary movement in as many as four axes are used, especially for perforation by EBM of jet engine components.

Foil made of a synthetic material has been perforated with 620 holes per

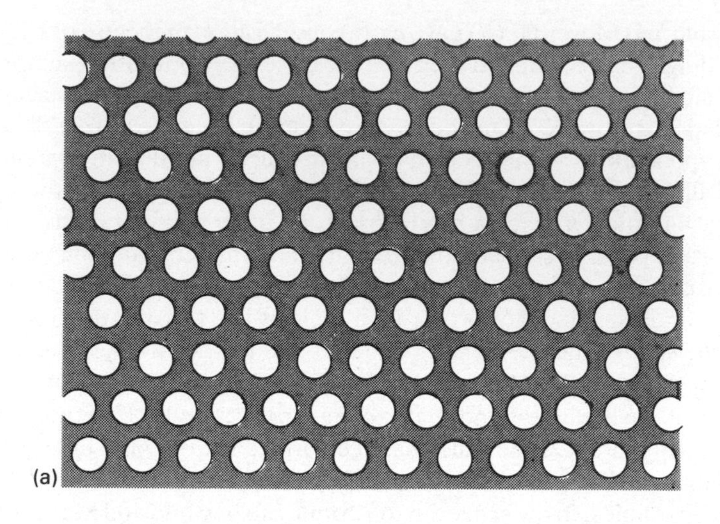

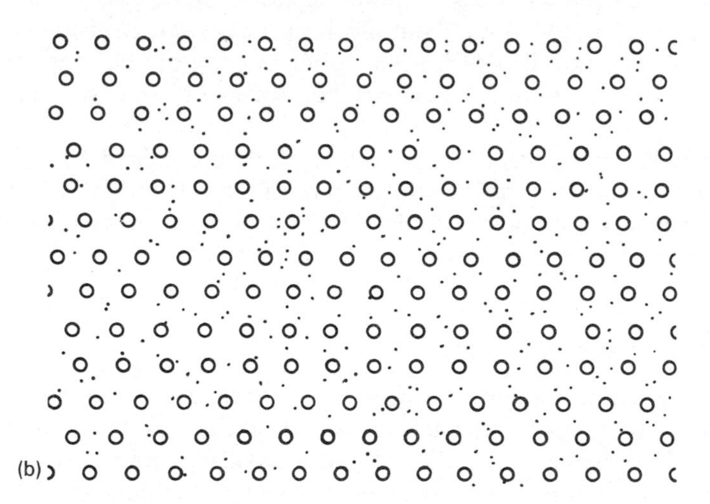

Fig. 2.13 Pattern of holes drilled by EBM. (After Steigerwald and Meyer, 1967.)

(a) Workpiece material: stainless steel; thickness: 0.2 mm; diameter of holes: 0.09 mm; density of holes: 4000 per cm^2; distance between holes: 0.16 mm; distance between rows: 0.16 mm; time required to drill one hole: 10 μs.

(b) Workpiece material: synthetic fabric; thickness: 0.012 mm; diameter of holes: 0.006 mm; density of holes: 20 000 per cm^2; distance between holes: 0.07 mm; distance between rows: 0.07 mm; time required to drill one hole: 2 μs.

square millimetre for filter applications at a rate of one hole every 10 μs. Some melting does occur, especially with plastics, as occurred with the fabric shown in Fig. 2.13(b) where foil with 205 perforated holes per square millimetre was produced.

Finally, Fig. 2.15 shows a cross-section of a hole 0.125 mm in diameter drilled in 30 μs through a sheet of nickel alloy 0.4 mm thick. This technique can be applied to the production of filters, and masks for colour television tubes. Other applications for perforation lie in sieve manufacture, for sound insulation and in glass fibre production.

2.8.3 Pattern generation for integrated circuit fabrication

Birnie and Champney (1967) were amongst the first to draw attention to the use of electron beam technology in scribing thin film circuits for the electronics industry. An attraction of the former is the wavelengths that are some orders of magnitude shorter than those of light systems, which were used before integrated circuits became so complex.

A detailed account of EBM for the manufacture of integrated circuits has been presented by Yew (1977). The beam is positioned accurately by means of deflection coils at the location where a pattern is to be written, by exposing a film of electron resist coated on either a chrome mask blank or a wafer, for the production of the lithographic definition required.

Electron resists may be usefully explained here. These are polymeric materials, similar to photo resists used in lithography, except that the former are sensitive to exposure to electrons rather than ultraviolet light (as is the case with the latter).

An electron beam of energy about 10 to 20 kV can readily break the bonds between polymer molecules, in the case of positively acting electron resists. It can also cause cross-linking in the polymers, for negatively acting electron resists. With the onset of either of these conditions, the solubility changes when the resist film is immersed in the developer, usually a solvent for the resist. Due to the difference in solubility between the original and exposed resist polymers, differential material removal occurs. A fine pattern of polymer is thus obtained. This pattern is then used as an active mask to avoid unwanted etching of the integrated circuit mask or wafer.

Since integrated circuits are usually produced on silicon wafers, 75 mm or greater in diameter, a moving work table is employed to position precisely each area of the chip under the electron beam, in order that the required pattern can be produced. The accuracy of this operation relies greatly on control of the relative position between the electron beam and the substrate. The two main systems used, field-to-field registration with bench marks and laser interferometry, have been discussed by Yew (1977).

Table 2.2 Applications of EBM (after Birnie and Champney, 1967)

Material	*Voltage (kV)*	*Current (μA)*	*PW* μs*	*PF† (cps) (Hz)*	*Beam focus condition*	*General comments*
Alumina wafers						
0.25 mm gauge	90	150	80	150	Focused	0.10 mm slot cut at 300 mm min^{-1}
0.25 mm gauge	125	60	80	50	Focused	0.075 mm hole cut in 10 s
0.75 mm gauge	150	200	80	200	Focused	0.10 mm slot cut at 610 mm min^{-1}
0.75 mm gauge	125	60	80	50	Circle generator	0.30 mm hole cut in 30 s
Sapphire crystal						
0.65 mm gauge	110	20	9	50	Beam circularly	0.064 mm hole – <30 s
Ferrite wafers						
(0.25 mm thick)	140	25	5	50	Focused	0.025 mm diameter holes drilled in <1 s
Molybdenum shim						
(0.25 mm thick)	140	20	20	50	Focused	<0.050 mm diameter holes drilled in <1 s on 0.075 mm centres

Microdiodes						
Scribing	110	7	12	50	Focused	Scribed to ≈0.025 mm depth. Scribing rate ≈1500 mm min^{-1}. (Maximum rate not evaluated)
Silicon wafers						
(0.25 mm thick) (gold deposited)	130	70	4	3000	Focused	Scribed to 0.05 mm depth at 127 mm min^{-1}
Thin film register						
(Tantalum, 100A)	100	20	9	1000	Focused	Cut by manual programming
Myler tape						
(0.038 mm)	110	600	Continuous beam		Focused	117 200 mm min^{-1}
Steel drill						
(0.36 mm diameter)	140	200	80	50	Focused circular deflected	Drilled in ≈3 min

*Pulse width
†Pulse frequency

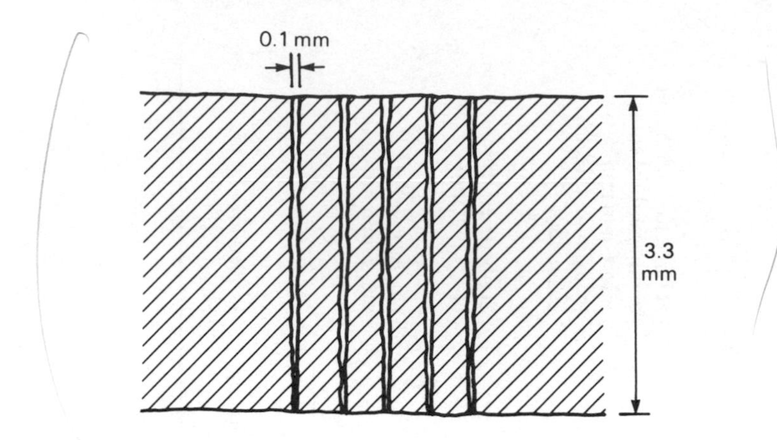

Fig. 2.14 Cross-section of holes drilled by EBM. (Workpiece material: high-temperature nickel alloy.) (After Steigerwald and Meyer, 1967.)

Table 2.3 Drilling of nickel alloy (accelerating voltage 130–150 kV) (Data from Steigerwald and Meyer, 1967)

	(a)	*(b)*	*(c)*	*(d)*
Thickness (mm)	3.3	3.3	3.3	1.6
Length of drilled hole (mm)	3.3	5.2	5.2	3.1
Diameter of drilled hole (mm)	0.1	0.7	0.2	0.3
Taper (°)	0	2	1	0
Angle of drilling (°)	90	35	35	30
Time of drilling (μs)	1	15	3	10

The pattern generation is carried out by vector or raster scanning. With the former technique depicted in Fig. 2.16, the electron beam is deflected only to locations at which the electron resist is to be exposed. As soon as the deflection system completes the positioning of the beam at the required place, the electron beam action is started. The vector scan method is particularly attractive for low-density patterns because of the amount of time that it saves.

As illustrated in Fig. 2.17 with the raster scan system, the chip pattern is first divided into subfields. Each subfield is scanned by the electron beam in a raster, like that employed with television. The electron beam is turned on and off along each raster line as needed. The required pattern is fully formed, by the combined effects of electron beam exposure and subsequent resist development. With the raster scan method, the electron beam has to cover

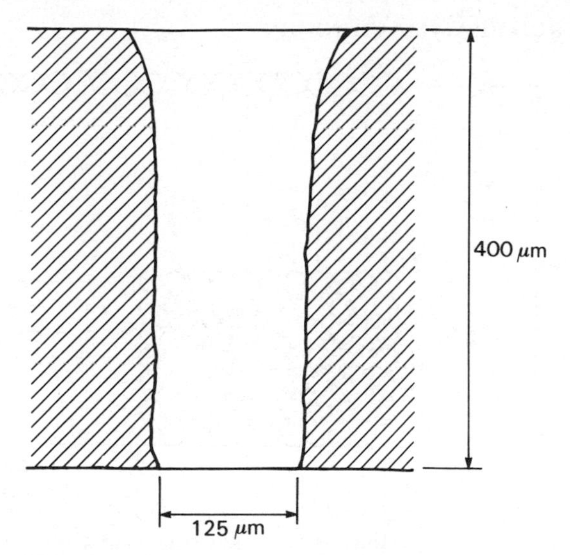

Fig. 2.15 Hole drilled in metal alloy sheet.

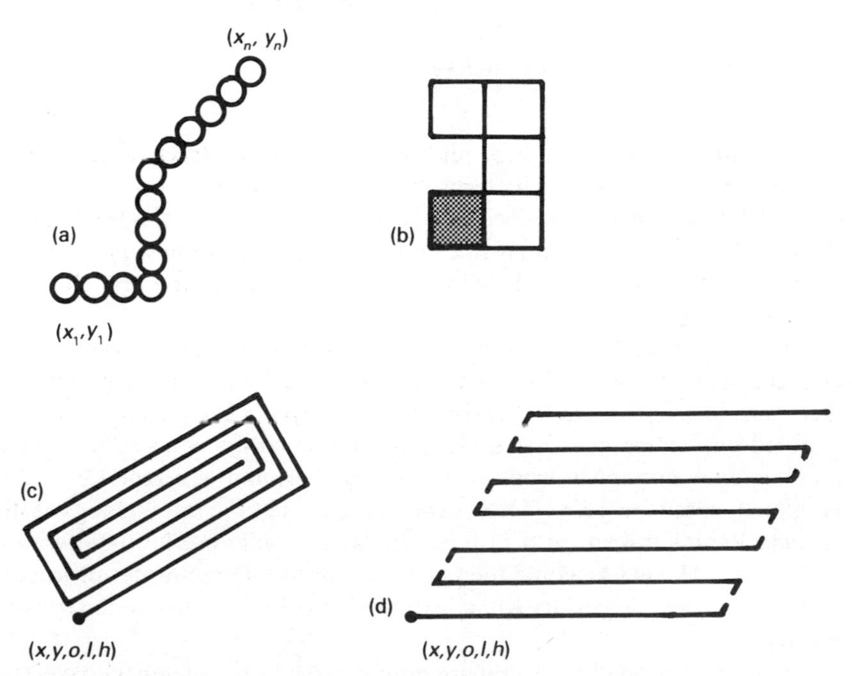

Fig. 2.16 Vector scan writing method. (After Yew, 1977.)
(a) Simple. Each dot has (x, y) address output from a computer through a D/A converter to a beam deflection and blanking system.
(b) One spot: 2.5 μm × 2.5 μm. Minimum positioning step: 0.25 μm.
(c), (d) Deflected beam: chip 6 × 6 mm (max).

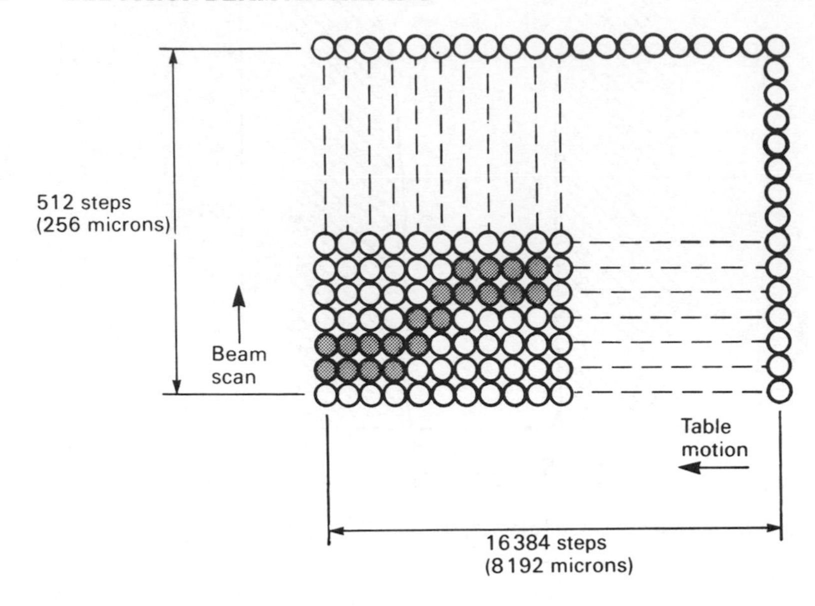

Fig. 2.17 Raster scan method. (After Yew, 1977.)

most of the mask or wafer area and is therefore less attractive than vector scanning, when low density patterns have to be produced.

The EBM system can produce primary chrome masks in a single step, since the electron beam is extremely fine and can have its position very accurately controlled. With such a single step method there is of course less chance of defects arising.

When the electron beam is used to manufacture primary chrome masks, conventional chrome and resist-coated glass blanks are used, except that electron, instead of photo, resists are employed. The electron resist is exposed until complete exposure is achieved on all the chips. The exposed blank is then withdrawn from the work chamber, and developed by a conventional spray process. After development subsequent etching of the chrome is undertaken, in a fashion similar to common photolithographic techniques. The mask should then be ready for use. Direct wafer processing by EBM becomes necessary when feature sizes below one to two microns are desired.

Line widths as small as several hundred Å can now be written with electron beam techniques, and writing speeds up to about 20 MHz are obtainable. Across a 125 mm mask an accuracy of 0.125 μm can be achieved, and a mask of this size can be manufactured in about 60 minutes. Direct wafer processing electron-beam systems have been built to produce up to 22 wafers per hour.

In more recent reports on electron beam lithography, Richman (1987) describes a machine capable of producing masks or wafer levels up to 150 mm across, with a 0.125 μm address size, and figure placement accuracies better than 0.02 μm. He comments that industrial requirements for machines capable of 0.25 μm feature size, at ten or more wafer levels per hour will probably be achieved in the 1990s.

BIBLIOGRAPHY

Bellows, G. (1976) *Non-traditional Machining Guide – 26 Newcomers for Production*, Metcut Research Associates Inc., Cincinnati, Ohio, pp. 40, 41.

Bellows, G. and Kohls, J. B. (1982) Drilling without Drills, Special Report 743, *American Machinist*, March, pp. 185, 186.

Birnie, J. V. and Champney, M. A. (1967) The Contribution of Electron Beams to Machining and Forming, in *Electrical Methods of Machining and Forming*, IEE Conf. Publ. no. 38, pp. 210–21.

Boehme, D. (1983) *Perforation Welding and Surface Treatment with Electron and Laser Beam*, Proc. 7th Int. Symp. on Electromachining IFS (Publications) Ltd, pp. 189–200.

Dugdale, R. A. (1970) Soft Vacuum Electron Beam Processing with Glow Discharge Guns, in *Electrical Methods of Machining, Forming and Coating*, IEE Conf. Publ. no. 61, pp. 48–53.

Einstein, P. A. and Beadle, R. (1965) *Electron-Beam Operation on Materials*, Conf. on Machinability, London 4–6 October, published in 1967 by the Iron and Steel Institute, pp. 227–33.

Kaczmarek, J. (1976) *Principles of Machining by Cutting, Abrasion and Erosion*, Peter Peregrinus Ltd, Stevenage, pp. 514–28.

Richman, R. M. (1987) *Precision Engineering Issues in Electron Beam Lithography Machines*, Proc. 3rd International Precision Engineering Seminar, Cranfield.

Steigerwald, K. H. and Meyer, E. (1967) *New Developments in Electron Beam Machining Methods*, Electrical Methods of Machining and Forming, Institution of Electrical Engineers, Conf. Publ. no. 38, pp. 252–8.

Yew, N. C. (1977) *Electron Beam – Now a Practical LSI Production Tool*, *Solid State Technology*, August, pp. 86–90.

3 Ion beam machining

3.1 INTRODUCTION

Ion beam machining takes place in a vacuum chamber, with charged atoms (ions) fired from an ion source towards a target (the workpiece) by means of an accelerating voltage. The process works therefore on principles similar to electron beam machining, although, as will be made clear later in the chapter, the mechanism of material removal is quite different.

Ion beam machining (IBM) is closely associated with the phenomenon of 'sputtering', first reported by Grove in 1852 (Carter and Colligon, 1968). Whilst investigating the electrical conductivity of gases, he discovered that metallic substances had become deposited on the glass walls of the glow discharge tube which he was using. Grove inferred that metal atoms had been removed from the surfaces of the electrode, and subsequently had adhered to the walls of the glass tube. Later the mechanism underlying Grove's finding was established as the ejection of atoms from a surface when it is bombarded by other ions.

3.2 ION BEAM MACHINING SYSTEM

An ion beam machine has three main components:

(1) a plasma source which generates the ions;
(2) extraction grids for removing the ions from the plasma, and accelerating them towards the substrate (or specimen);
(3) a table for holding the specimen.

3.2.1 Ion source

For the removal of an atom from the surface by impingement of an ion, or ions, a source of ions is required which should produce a sufficiently intense beam, with an acceptable spread in its energy.

Jolly, Clampitt and Reader (1983) have reviewed suitable ion sources for IBM equipment. They draw attention to the 'electron-bombardment' ion source, or Kaufman system. This ion source was developed as a comparatively low thrust motor for adjusting the altitude and orbits of satellites. Its main characteristics have been summarized by Spencer and Schmidt (1972) and are presented in Table 3.1, together with information on the 'duoplasmatron' source which is also of interest for IBM.

Table 3.1 Characteristics of ion beam sources (Data from Spencer and Schmidt, 1972)

Ion gun	*Beam current (mA)*	*Beam current density ($mAcm^{-2}$)*	*Beam voltage (kV)*	*Beam diameter (cm)*
Kaufman	10–50	0.85 at 1 kV	0.5–2.0	5.0
Duoplasmatron	10	10^3	0–25	Focus to 0.3 mm

3.2.2 Plasma source

As indicated in Fig. 3.1, a heated filament, usually tungsten, acts as the cathode, from which electrons are accelerated by means of a high voltage, above 1 kV, towards the anode.

During the passage of the electrons from the cathode to the anode, they interact with argon atoms in the plasma source (which is sustained by keeping the gas pressure at about 10^{-4} torr). The following reaction then occurs:

$$Ar + e^- \rightarrow Ar^+ + 2e^-$$

Argon ions are thereby produced. A magnetic field, obtained from an electromagnetic coil or a permanent magnet, is often applied between the anode and cathode to make the electrons spiral. Spiralling increases the path length of the electrons and hence increases ionization.

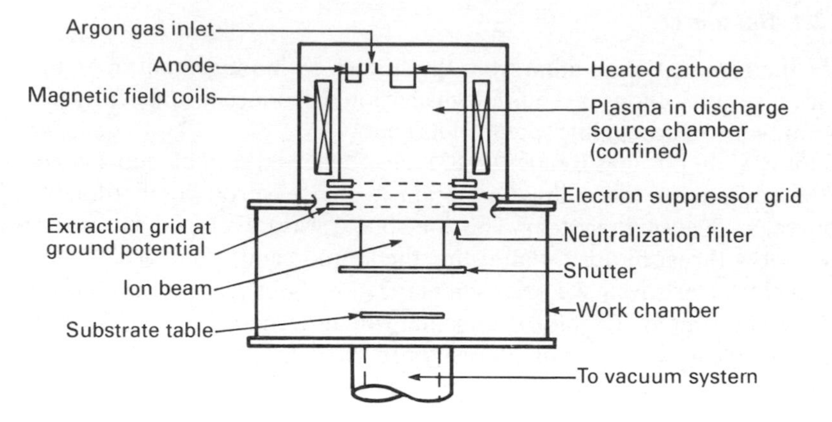

Fig. 3.1 Main features of ion beam machine. (After Melliar-Smith, 1976.)

3.2.3 Extraction grids

The ions are removed from the plasma by means of extraction grids. The grids are normally made of two or three arrays of perforated sheets of carbon or molybdenum; these materials can withstand erosion by ion bombardment. The perforations in each of the sheets are aligned above one another. Indeed the shape of the holes and the spacings of the grids are significant elements when ion source systems are being designed to give the best conditions of ion current and grid erosion.

The outer grid is usually kept at ground potential, which is a more negative level than that of the anode. This grid therefore provides the negative field that is needed to remove the ions from the plasma. The second grid is held at a negative potential below the ground value. The escape of electrons from the plasma is thereby prevented, as is their diffusion back from the work chamber. A third grid, which is maintained at the anode potential, is sometimes added – placed between the plasma and the electron suppressor grid – to improve the performance of the source.

Extraction voltages of 0.5 to 2 kV with associated current densities of approximately 2 $mAcm^{-2}$ are used with Kaufman ion sources.

3.2.4 Substrate mounting

When the ions have been removed from the source, they 'drift' in a field-free region to the component, specimen, or substrate, which is to be machined, or milled.

As shown in Fig. 3.1, the specimen is usually mounted on a water-cooled table which can be tilted through an angle of 0 to 90°. The specimen is separate from the plasma. Machining variables such as acceleration, flux and angle of incidence can all be independently controlled.

The current I which is associated with the extraction of the ion may be calculated from Child's law:

$$I = \frac{\pi \epsilon_0}{q} V^{3/2} \left(\frac{2q}{m}\right)^{1/2} \left(\frac{d}{l_e}\right)^2 \tag{3.1}$$

where ϵ_0 is the dielectric permittivity, V is the potential difference between the grids, q and m are respectively the charge and mass of the ion, d is the diameter of the apertures in the grid, and l_e is the effective separation of the grid.

From equation (3.1) it is noted that the highest currents are obtained at the lowest spacing between the grids, and for grids carrying the largest number of holes of the smallest size.

In Fig. 3.2 an ion source from which is produced a 25 mm diameter beam of argon (Ar^+) ions, of current 110 mA is shown striking an alumina target

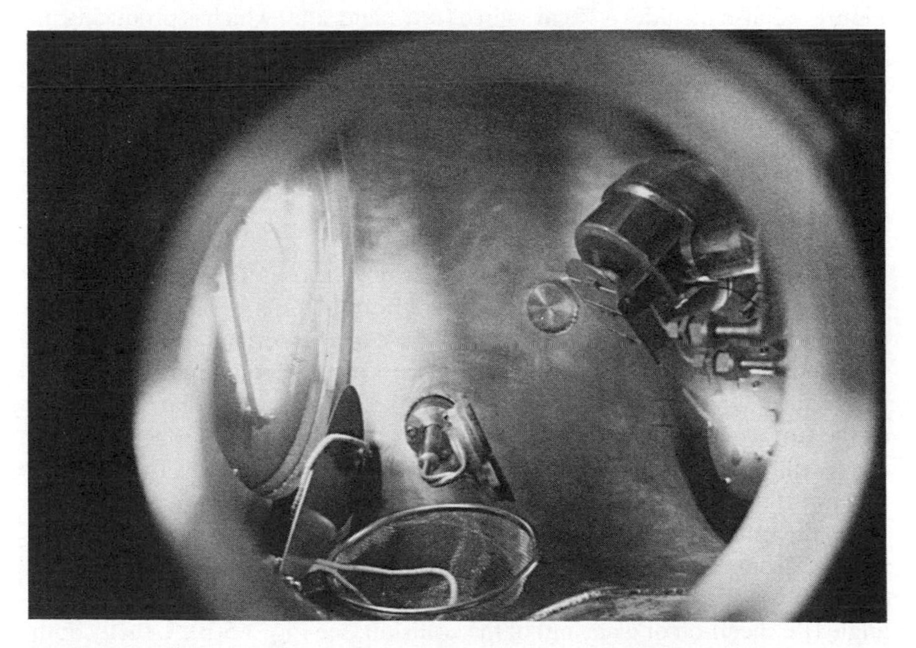

Fig. 3.2 Argon ion beam striking alumina target. (Courtesy of Oxford Applied Research.)

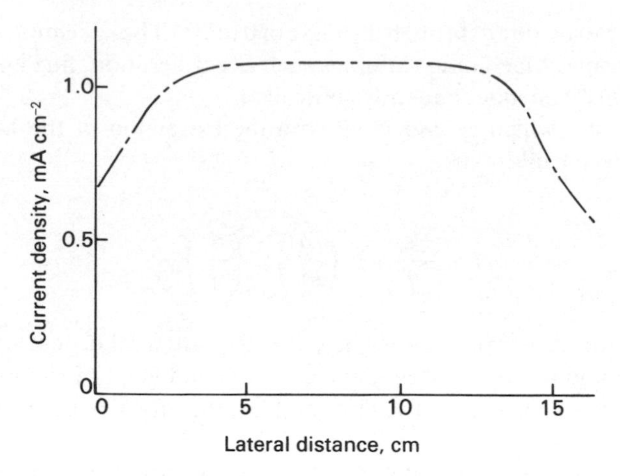

Fig. 3.3 Profile of ion beam from 15 cm source. (After Jolly, Clampitt and Reader, 1983.) Source-probe distance: 10 cm; beam energy: 500 eV; beam current: 202 mA; gas flow (STP): 2.4 ml min^{-1}.

(which is in the lower centre of the photograph). Surfaces mounted on a water-cooled rotating holder are coated with the sputtered alumina (top right). Figure 3.2 also includes a 15 cm source (left-hand side) which is producing up to 0.5 A of oxygen ions at 500 eV to mill the surface of the alumina specimen. Figure 3.3 shows the profile of the beam from this source.

An industrial version of an ion beam machining system is shown in Fig. 3.4.

3.3 COLLISION MECHANISM

The interactions of the ions with atoms may now be usefully considered.

Firstly, it is noted that the size of an ion is normally comparable with that of an atom. Thus when an ion strikes the surface of a material it usually collides with an atom there. This collision often occurs in a direction that is normal to the surface. On the one hand, if the mass of the ion is less than that of the atom of the surface, the former will 'bounce' back, away from the surface; and the atom will be driven in a direction further into the material. Figure 3.5(a) illustrates this condition.

On the other hand, if the mass of the incident ion is greater than that of the surface atom, after collision, the ion and atom move from the position of collision towards the interior of the surface of material, irrespective of the angle (i.e. head-on or glancing) of the collision, see Fig. 3.5(b). Usually both particles move into the material at energies that are less than that of the incident ion, yet much greater than the lattice energy.

This type of primary collision, as it is called, does not lead to removal of atoms from the surface.

An alternative set of circumstances arises if the ion strikes the atom at a glancing angle. One condition, shown in Fig. 3.6(a), involves the incident ion impinging on a stationary surface atom at an angle of 90°. This atom does not obtain a velocity component in a direction away from the surface, as a direct consequence of this primary collision.

However the primary collision will cause at least one, and often two, further secondary 'binary' collisions just below, and very close to, the surface. Figure 3.6(b) illustrates this case. After the collision between ion and atom either particle should be able to leave the point of impact at more than 45° to the original direction of velocity of the ion.

Then, a secondary collision should be possible in the same plane of motion, to result in a lattice atom leaving the point of secondary impact at an angle greater than 45°. That is, a total angle of more than 90° is obtained. With an angle of this size involved the lattice atom has a velocity component which is in a direction outward from the surface. The atom has therefore the capacity to be ejected.

However the atom cannot be ejected in a direction parallel to the normal to the surface, that is, in a direction opposite to that of the incident ion. That movement would require two 90° deflections. At least one of these deflections would involve the deflection of a lattice atom through 90°, during which it would acquire zero velocity. An atom with zero velocity cannot be ejected. For the same reason the atom cannot cause ejection.

Although ejection might then be expected to occur most commonly in directions away from the normal to the surface, experimental observations have shown that for ion impingement on the normal direction to the surface, atoms are ejected from the surface in a cosine distribution. That is, ejection is most likely to occur in a direction that is exactly the opposite to that of the incident ion (it should be noted that this occurs only for ion bombardment at normal incidence). Apparently the energy imparted by the incident ion is so randomly distributed by multiple collisions before the atom is ejected that the incident momentum vector is fully lost. It then has no influence on the process of ejection.

If the incident ion strikes the surface obliquely, the ejection is very likely to result from the primary collision between the incident ion and the surface atom with which it first collides. In this case experiments show that the incident momentum vector has a great influence on the ejection process. The atoms are found to be ejected mainly in the forward direction.

The sputtering yield, that is, the number of atoms ejected per incident ion, may be as much as an order of magnitude greater for oblique, rather than normal, incidence.

Typical specifications

Beam Diameter	25mm	110mm	150mm
Current density @ voltage mA/cm^2	1 @ 250eV 10 @ 1500eV	0.8 @ 100eV 2 @ 500eV 4 @ 1500eV	2.5 @ 1500eV
Uniformity	± 10% over 12mm	± 5% over 80mm	± 5% over 120mm
Ionization efficiency	85%	85%	85%

Fig. 3.4 Elements of industrial ion beam machine. (a) Upper part of photograph shows vacuum chamber with specimen holder on the right. Left-hand side of photograph shows control cabinet and power system. Main elements are shown in (b), (c) and (d). (Courtesy of Oxford Instruments Ltd.)

(b) 150mm Ion Source

(c) Specimen Holder

(d) Sputter Target Holder

Spencer and Schmidt (1972) have also considered the mechanisms at work in IBM. They explain material removal in terms of the transfer of momentum from the incident ions to atoms on the surface of the material. Thus in Fig. 3.7 an atom is removed from the surface, and the ion is also deflected away from the material. Spencer and Schmidt propose that energies greater than the binding energy of 5 to 10 eV are needed to effect removal of atoms.

As indicated also in Fig. 3.7, at higher energies, sufficient momentum may be transferred by the impinging ions for several atoms to be removed from the material, in a 'cascade-type' effect near its surface.

The higher the energy of the incident ion, the more deeply this cascading effect occurs into the material. Several atoms, or molecules, ionized or

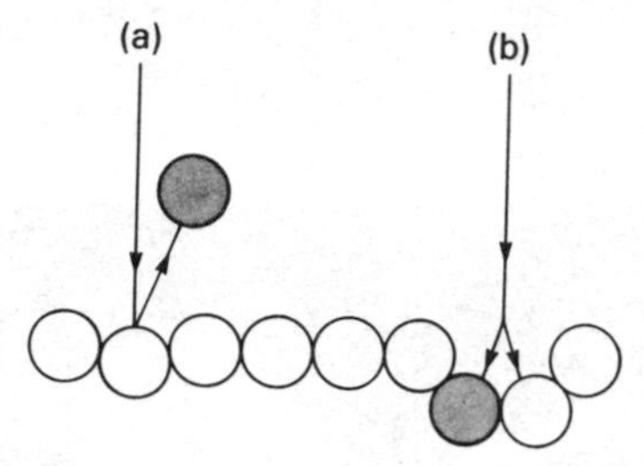

Fig. 3.5 Ion (shaded) bombardment at normal incidence to surface. (After Stuart, 1983.)

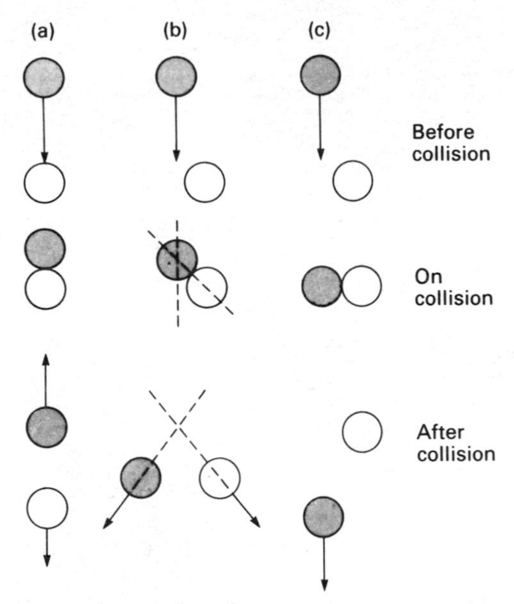

Fig. 3.6 Three types of collision between ion (shaded) and atom (mass of ion less than that of atom). (After Stuart, 1983.)

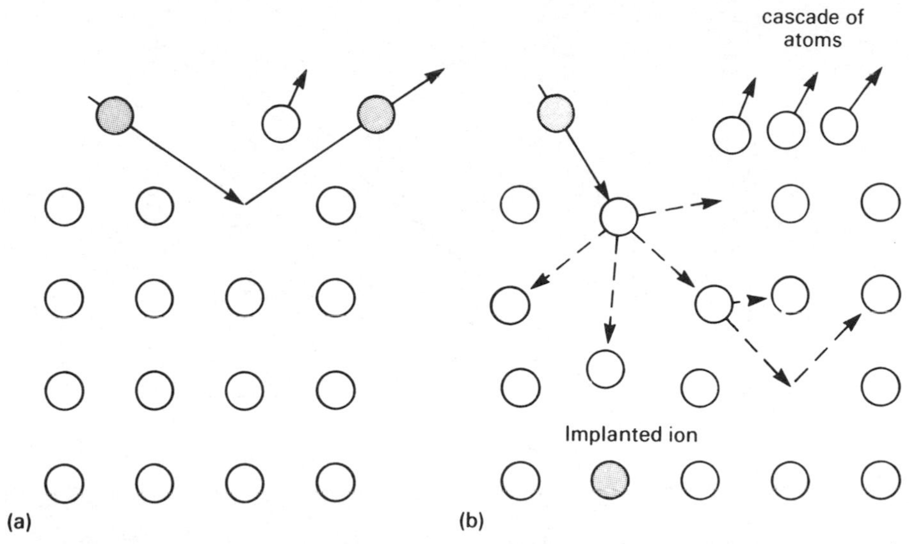

Fig. 3.7 Effects of low and high energies on atom removal. (After Spencer and Schmidt, 1972.)
(a) Low energy case.
(b) High energy case.

neutral, are likely to be ejected from the material. The incident ion will become implanted deep into the material, damaging it, by displacement of atoms.

Various theoretical expressions have been reported for the yield, that is the number of atoms removed per incident ion. See for example Spencer and Schmidt (1972) and Somekh (1976). These workers all confirm that the yield depends on the material being treated, the type of atoms and their energy, the angle of incidence, and in some cases, the gas pressure. The relationships that are presented depend often in a complicated way on other fundamental quantities. Despite the undoubted values of these expressions, the more practical aspects of the process can be readily assessed from experimentally obtained values.

3.4 RATES OF MATERIAL REMOVAL IN IBM

Typical experimental results in which yield increases nonlinearly with incident ion energy, for argon ions impinging on a range of materials, are shown in Fig. 3.8. For higher ion energies a yield of approximately 0.1 to 10 atoms per incident ion is representative of IBM. Spencer and Schmidt discuss the effects on yield of a wide range of process conditions, not only incident ion

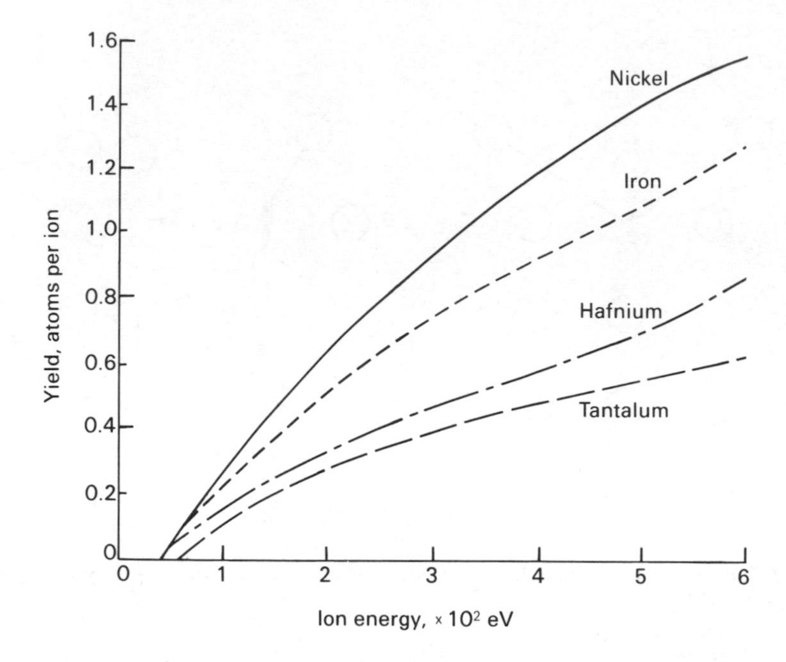

Fig. 3.8 Variation of yield with ion energy. (After Spencer and Schmidt, 1972.)

energy but also the angle of incidence, ionic and atomic periodicity, and heat of sublimation. They point out, for example, that the rate of material removal (or etching) can also be enhanced at grain boundaries and can be affected by crystallographic orientation.

An alternative way of recording rates of removal is in terms of milling rate, which typically varies between 1 and 2 μms^{-1} for, respectively, normal and 50° angles of incidence. Some typical milling rates for a range of materials which have been ion beam machined are presented in Table 3.2.

Useful experimental information is available on these results which supplements that given in Table 3.2. The results were obtained with apparatus which included a low voltage gun, utilizing a hot filament, as the source of electrons to ionize the argon gas. The hollow anode source used as the ion gun was about 5 cm in diameter and 15 cm long. The anode and cathode had matching arrays of 300 holes, each of 0.3 mm diameter, and spaced apart such that uniform material removal or milling, accurate to within about 3% over the machining area, could be obtained. With this anode–cathode configuration an area of 300 cm diameter could be treated. Each beam was capable of 100 μA with only 1 kV voltage applied; that is, the total beam current was 30 mA.

Table 3.2 Typical removal (milling) rates by IBM (hollow anode source) (Data from Spencer and Schmidt, 1972)
Data: argon ion beam 60 to 70° from normal
pressure = 3×10^{-4} Torr
voltage = 6 kV
current = 100 μA
current density = 1 mAcm^{-2} over 1 cm diameter area

Material	*Removal (milling) rate (μmhr^{-1})*
Quartz	2
Garnet	1
Ceramic	1
Glass	1
Gold	2
Silver	3
Photo resist material (KTFR)	1
Permalloy	1
Diamond	1
GaAs (500 μA current)	10
GaP (500 μA current)	125

Further experimental results for an argon ion beam at normal incidence are presented in Tables 3.3 and 3.4.

From all these results, other relevant data concerning IBM can be summarized. For a typical IBM operation on a material of area 1 cm^2, a depth of approximately 10^{-4} mm is removed with an ion beam of 100 μA current. At a removal rate of one atom per incident ion, 1 μmhr^{-1} of material would be machined over that area; that is about 3 atoms s^{-1} or one atomic monolayer per second per square centimetre of area. The removal rate can be increased by use of greater beam fluxes or higher energy ions, although lower energy conditions are preferable. For most practical circumstances, IBM can be regarded as a 'sequential' operation in which each ion can be considered to act alone. As single ions impinge on the material each one can be regarded as removing an atom.

An expression for etch-rate in terms of yield has been given by Somekh (1976). He confirms that the yield depends on the material being etched, the type of atoms and their energy, the angle of incidence and, in some cases, the gas pressure. Somekh proposes that the etch-rate $V(\theta)$ (atoms min^{-1} per mA cm^{-2} of incident flux) is related to the yield S by

$$V(\theta) = 9.6 \times 10^{25} \frac{S(\theta)}{n} \cos\theta \left(\frac{\text{atoms min}^{-1}}{\text{mA cm}^{-2}}\right) \tag{3.2}$$

Table 3.3 Removal rates by IBM (After Spencer and Schmidt, 1972)
Data: incident beam normal to surface
pressure = 3×10^{-4} torr
voltage = 1 kV
current = 0.85 $mAcm^{-2}$
beam diameter = 5 cm

Material	*Removal (milling) rate (μmhr^{-1})*
Silicon	2.0
Ga As	15.9
Silica (ceramic)	2.3
KTFR photoresist	2.3
Silver	18.0
Gold	2.6

Table 3.4 Removal (milling) rates obtained with 500 eV argon ions incident normally on target. Current density 1 $mAcm^{-2}$ (Data from Jolly *et al.*, 1983)

Material	*Removal (milling) rate nms^{-1}*
Carbon	0.07
Aluminium	1.2
Silicon	0.62
Chromium	0.83
Manganese	1.5
Silver	2.5
Gold	2.4
SiC	0.52
SiO_2	0.67
Fe_2O_3	0.78
AZ 1350 (photoresist)	0.50
PMMA (photoresist)	0.93

where n is the atomic density (atom cm^{-3}) of the target material; the '$\cos\theta$' term takes into account reduced current densities at angles away from normal incidence.

He points out the dependence of yield on the binding energy of the atoms in the material being etched. The amount of yield can therefore be varied, by the introduction of reactive gases.

These reactive gases can react with the surface of the material, vary its binding energy and hence the rate of material removal (i.e. the etch-rate). For example, a flux of reactive species containing fluorine is known to react with materials such as titanium or silicon to form loosely bound or volatile compounds, which increase the etch-rate.

Jolly *et al.* (1983) discuss further aspects of reactive IBM. They have investigated conditions in which a flux of a reactive species such as methane (CH_4), fluorine (CF_4) rather than inert argon ions is directed at the specimen-target.

The free radicals chosen should react with the material at its surface, forming volatile products or those which can be readily milled (removed) by the effect of the kinetic energy of the bombarding ions.

3.5 ACCURACY AND SURFACE EFFECTS

Jolly *et al.* (1983) claim that dimensions as small as 100 nm should be possible by IBM, reporting that features of size less than 10 nm were obtainable at that time. The slope of the walls of the machined surface and its surface finish are determined by the angle of incidence of the ion beam which is fully controllable. They report that the process variables can be monitored to an accuracy of ±1.0%, with a repeatability of ±1.0%.

Surfaces can be textured by IBM, which can produce cone- and ridge-like configurations, of the order of 1 μm in size. Smoothing of a surface to a finish of less than 1 μm can also be achieved by IBM. Jolly and co-workers also claim that a surface that is already smooth can undergo ion beam machining without significant increase in its roughness. They report typical results, in which a depth of 100 nm can be machined from a surface; an initial surface roughness of 1.5 nm was found to be altered by less than ±1 nm.

3.6 APPLICATIONS

3.6.1 Smoothing

The use of IBM for smoothing of laser mirrors and for modifying the thickness of thin films and membranes without affecting surface finish is reported by Jolly, Clampitt and Reader (1983).

3.6.2 Ion beam texturing

Hudson (1977) has demonstrated that an ion-beam source is a controlled method for texturing surfaces. A typical result is presented in Fig. 3.9 in which a structure resembling closely packed cones was produced. As well as the nickel and copper shown in Fig. 3.9 Hudson went on to investigate 26

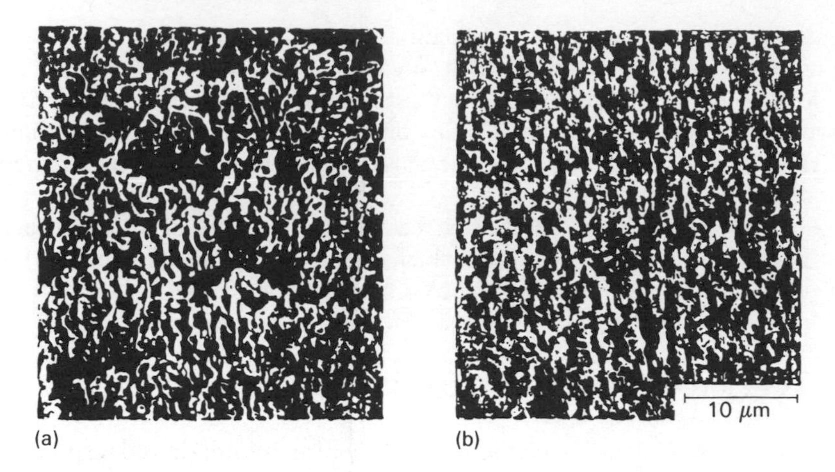

Fig. 3.9 Scanning electron photomicrographs. (After Hudson, 1977.)
(a) Nickel.
(b) Copper.

materials, including stainless steel, silver and gold. He has also summarized a number of other investigations in which a microscopic surface texture is created by sputter-etching performed simultaneously with the sputter-deposition of a lower yield material onto the surface.

Applications of ion beam texturing have been discussed further by Jolly *et al.* They include enhanced bonding of surfaces, increased surface areas of capacitors, and surface treatment of medical implants.

3.6.3 Ion beam cleaning

Atomically clean surfaces can be produced by IBM. This technique can be preferable to electron beam and electrical discharge methods which can damage the surface. Harper, Cuomo and Kaufman (1982) discuss in detail this well-established application of ion beam technology. For example, they report substantial improvements in the adhesion of gold films to silicon and aluminium oxide Al_2O_3 substrates by use of argon or oxygen ion beam sputter-cleaning of the substrate, prior to evaporation. The cleaning consisted mainly of removal of absorbed water and hydrocarbons. When a layer of surface oxide has to be removed in order to clean a surface, higher ion energies, of several hundred eV, are needed. Damage to the substrate material may then arise: in this case a reactive gas with high selectivity of oxide etching can be used.

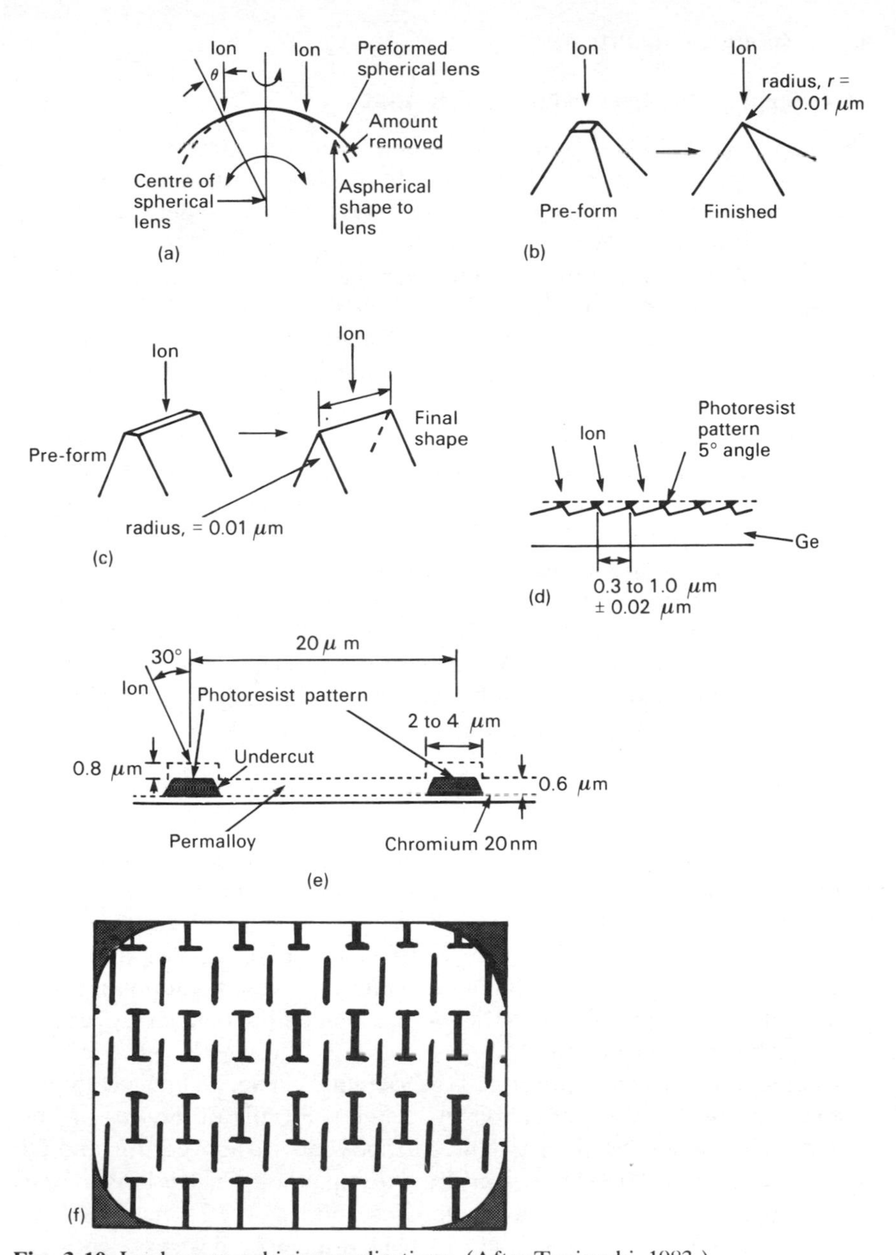

Fig. 3.10 Ion beam machining applications. (After Taniguchi, 1983.)
(a) IBM of aspheric lens.
(b) Sharpening of diamond indentor by IBM.
(c) Sharpening of diamond microtome cutter.
(d) Manufacture of holographic mask.
(e) Production of magnetic bubble memory.
(f) Ion milling of permalloy structure (10 div = 25 μm).

3.6.4 Shaping, polishing and thinning by IBM

Thinning by use of oblique incidence argon ions has been used to enhance polishing (Harper, Cuomo, Kaufman, 1982). Macroscopic thinning and shaping of materials can be applied to the fabrication of magnetic heads and surface acoustic wave devices.

The polishing and figuring of optical surfaces has also been reported by Harper, Cuomo and Kaufman (1982) and by Taniguchi (1983). The latter gives a valuable summary of the various applications of IBM including aspherizing of lenses, sharpening of diamond indentors and cutters and cutting tools. See Fig. 3.10.

Taniguchi (1983) points out that these operations are performed by the direct sputtering of pre-forms in glass, silica and diamond. Unlike conventional technology involving cutting, grinding, lapping and polishing, the ion beam process has no inherent polishing surface, e.g. guideways, the reference being the preform or patterning mask.

Thinning of samples of silicon to a thickness of 10 to 15 μm has been obtained by argon ions impinging at normal incidence. The production of samples for transmission electron microscopy (TEM) is another widespread practice; two opposing beams thin a circular region on a rotating sample until the centre etches through, leaving thin fringe areas suitable for TEM.

3.6.5 Ion milling

Jolly and co-workers (1983) report that ion milling is especially useful for the accurate production of shallow grooves, such as that illustrated in Fig. 3.11. Milling through masks, to produce regular arrays of pits with widths of 5 to 200 μm and depths of up to 1 mm for enhanced bonding, has also been achieved. These authors also point out that pillar-like configurations useful in the manufacture of precision electrical resistive and fibre optic arrays can be produced by ion beam methods.

Other workers have also confirmed the usefulness of ion milling techniques as an alternative to the fabrication by chemical etching of devices of fine geometry. The latter technique is limited to line widths of above 2 μm, and 1:1 depth-to-width ratios. IBM offers an alternative method which is limited only by masking capabilities. For example, line widths of 0.2 μm have been achieved by IBM in the fabrication of bubble memory devices; depth-to-width ratios 2:1 have been achieved. Problems associated with chemical etching, such as lack of line delineation owing to failure of resist adhesion and undercutting of layers, are avoided since masking is only needed to shadow the beam (Bollinger, 1977). Typical results from Bollinger's detailed account of the manufacture of solid state devices of fine geometry are given in Figs 3.12 and 3.13.

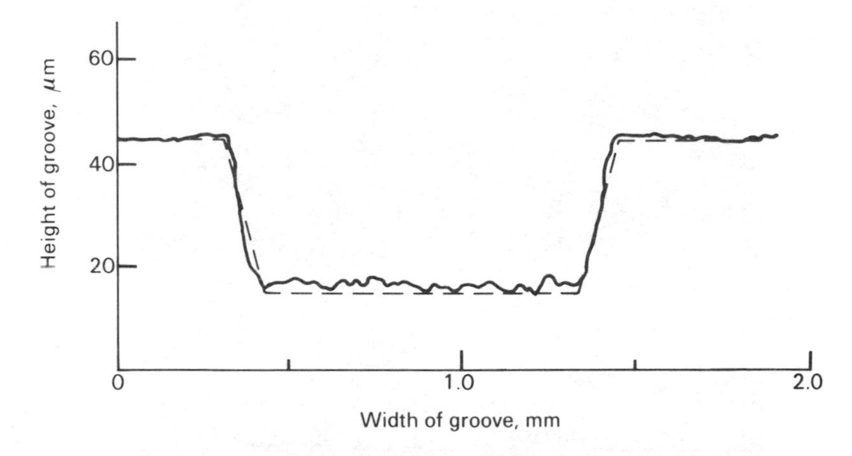

Fig. 3.11 Groove machined in refractory material. (After Jolly, Clampitt and Reader, 1983.) —— Profile obtained. ----- Required profile.

Fig. 3.12 Showing the ability of ion milling to etch near-vertical walls. (After Bollinger, 1977.)

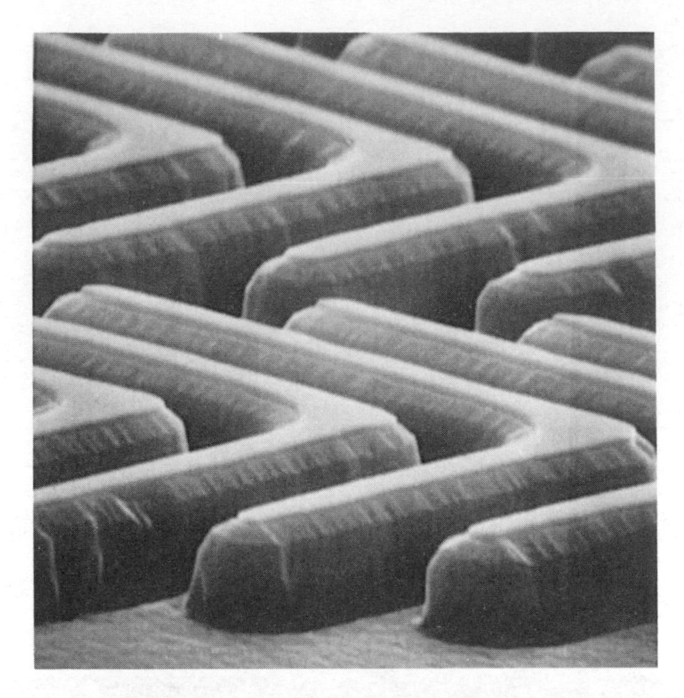

Fig. 3.13 A Ni–Fe 1 bar bubble memory pattern. (After Bollinger, 1977.)

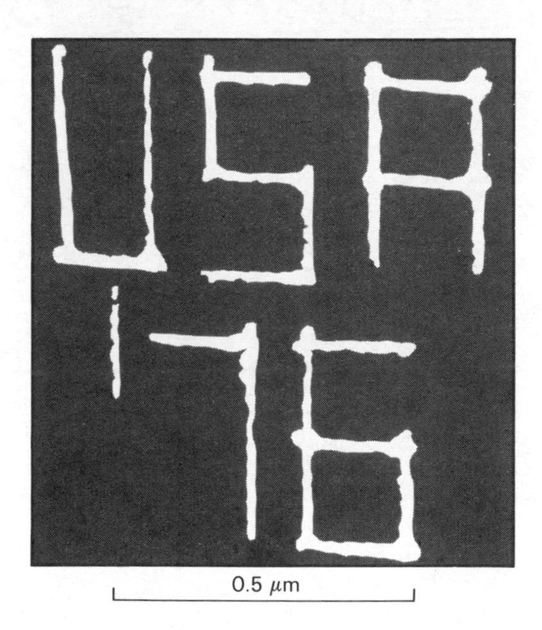

Fig. 3.14 Narrow line widths produced in carbon membrane by IBM. (After Harper, Cuomo and Kaufman, 1982.)

Figure 3.12 shows the fine etching by IBM of near vertical walls on a GaAs substrate. The milled gold structure accurately holds the pattern. No undercutting occurs at the gold-photoresist interface and the channel floor is flat right up to the wall interface. Figure 3.13 presents a nickel–iron (Ni–Fe) 1 bar bubble memory pattern etched by IBM with two μm line widths. The films are 0.5 μm thick, and the side walls produced are 10° off the vertical.

Further discussion on patterned microfabrication by IBM is given by Harper, Cuomo and Kaufman (1982). They describe the fabrication of line widths as narrow as 80 Å in 200 Å thick carbon membranes by argon ion beam etching (Fig. 3.14).

BIBLIOGRAPHY

Baudrant, A., Passerat, A. and Bollinger, L. D. (1983) Reactive Ion Beam Etching of Tantalum Silicide for VLSI Applications, *Solid State Technology*, **26** (9), 183–7.

Bollinger, L. D. (1977) Ion Milling for Semiconductor Production Processes, *Solid State Technology*, **20** (11), 66.

Bollinger, L. D. (1983) Ion Beam Etching with Reactive Gases, *Solid State Technology*, **26** (1), 99–108.

Bondur, J. A. (1976) Dry Process Technology (Reactive Ion Etching), *J. Vac. Sci. Technology*, **13** (5), 1023–9.

Carter, G. and Colligon, J. S. (1968) *Ion Bombardment of Solids*, Heinemann Educational Books Ltd, London, Chaps 7 and 9.

Dieumegard, D. (1980) Pulverisation et technologies d'érosion ionique, *Le Vide Les Couchas Minces*, **204**, 317–36.

Dimigen, H. and Luthje, H. (1975) An Investigation of Ion Etching, *Philips Tech. Rev.*, **35** (7/8), 199–208.

Doughty, G. F., Thoms, S., Law, V. and Wilkinson, C. D. W. (1986) Dry Etching of Indium Phosphide, *Vacuum* (Pergamon Journals Ltd), **36** (11/12), 803–6.

Garvin, H. L. (1973) High Resolution Fabrication by Ion Beam Sputtering, *Solid State Technology*, **16** (11), 31–6.

Gloersen, G. Per (1975) Ion Beam Etching, *J. Vac. Sci. Technology*, **12** (1), 28–35.

Gokan, H., Itoh, M. and Esho, S. (1984) Oxygen Ion Beam Etching For Pattern Transfer, *Vac. Sci. Technology*, **2**, 34–7.

Harper, J. M. E., Cuomo, J. J. and Kaufman, H. R. (1982) Technology and Applications of Broad-Beam Ion Sources Used in Sputtering, Part II Applications, *J. Vac. Sci. Technology*, **21** (3), 737–56.

Harper, J. M. E., Yeh, H. L. and Grebe, K. R. (1982) Ion Beam Joining Techniques, *J. Vac. Sci. Technology*, **20** (3), 359–63.

Hawkins, D. T. (1975) Ion Milling (Ion-Beam Etching) 1954–1975: A Bibliography, *J. Vac. Sci. Technology*, **12** (6), 1389–98.

Hudson, W. R. (1977) Ion Beam Texturing, *J. Vac. Sci. Technology*, **14**, 286–9.

Jolly, T. W., Clampitt, R. and Reader, P. (1983) Ion Beam Machines and Applications, *Proceedings of the 7th International Symposium on Electromachining*, Birmingham, 12–14 April 1983, pp. 201–10.

Kaczmarek, J. (1976) *Principles of Machining by Cutting Abrasion and Erosion*, Peter Peregrinus Ltd, Stevenage, pp. 528–30.

Kaufman, H. R., Cuomo, J. J. and Harper, J. M. E. (1982) Technology and Applications of Broad-Beam Ion Sources Used in Sputtering, Part I Ion Source Technology, *J. Vac. Sci. Technology*, **21** (3), 725–36.

Kaufman, H. R., Harper, J. M. E. and Cuomo, J. J. (1982) Developments in Broad-Beam Ion-Source Technology and Applications, *J. Vac. Sci. Technology*, **21** (3), 764–7.

McLeod, P. S. (1983) Reactive Sputtering, *Solid State Technology*, **26** (10), 207–11.

McNeil, J. R. and Hermann, W. C. Jr (1982) Ion Beam Applications for Precision Infra-Red Optics, *J. Vac. Sci. Technology*, **20** (3), 324–6.

Melliar-Smith, C. M. (1976) Ion Etching for Pattern Delineation, *J. Vac. Sci. Technology*, **13** (5), 1008–22.

Miyamoto, I. (1983) Ion Sputter-Forming of Knife-Edge Shaped Diamond Tools – for Over 90° Apex Angle, *Bull. Jpn Soc. Precision Engineers*, **17** (3), 195–6.

Nakai, T. *et al.* (1976) Application of Ion Beam Technique to Micro-Machining, *Bull. Japan Soc. of Prec. Eng.*, **9** (5), 157–62.

Somekh, S. (1976) Introduction to Ion and Plasma Etching, *J. Vac. Sci. Technology*, **13** (5), 1003–7.

Southern, A. L., Willis, W. R. and Robinson, M. T. (1963) Sputtering Experiments with 1 to 5 keV Ions, *J. Appl. Phys.*, **34**, 159–63.

Spencer, E. G. and Schmidt, P. H. (1972) Ion Beam Techniques for Device Fabrication, *J. Vac. Sci. Technology*, **8** (5), S52–S70.

Steinbruchel, C. H. (1984) Reactive Ion Beam Etching, *J. Vac. Sci. Technology*, **2**, 38–44.

Stuart, R. V. (1983) *Vacuum Technology Thin Films and Sputtering – An Introduction*, Academic Press, London, pp. 92–7.

Taniguchi, Norio (1983) Current Status in and Future Trends of Ultraprecision Machining and Ultrafine Materials Processing, *Annals of the CIRP*, **32** (2), 1–8.

Vossen, J. L. and O'Neil, J. J. Jr (1968) R-F Sputtering Processes, *RCA Review*, **29**, 149–79.

Wagner, A. (1983) Applications of Focused Ion Beams to Microlithography, *Solid State Technology*, **26** (5), 97–103.

Williams, M. M. R. (1981) The Energy Spectrum of Sputtered Atoms II, *Philosophical Magazine A*, **43** (5), 1221–52.

4 Electrochemical machining

4.1 INTRODUCTION

Another machining process which relies on the removal of atoms from the surface of a workpiece is electrochemical machining (ECM). Unlike ion beam machining (IBM), in which, as has been noted in the previous chapter, atoms are removed by transfer of momentum from the impinging ions as they strike the workpiece surface, ECM is an electrolytic process, with metal removal being achieved by electrochemical dissolution of an anodically polarized workpiece, one part of an electrolytic cell.

The first patent on a process resembling ECM was filed by Gusseff in 1929. However the first significant developments occurred in the 1950s, when ECM was investigated as a method for shaping high-strength, heat-resistant alloys which were difficult to cut by established methods. ECM is now employed in many ways, by the car, offshore and medical engineering industries, as well as by aerospace firms, its first main user.

4.2 ELECTROLYSIS

Electrolysis is the name given to the chemical process which occurs, for example, when an electric current is passed between two conductors dipped into a liquid solution. A popular application of electrolysis is the electroplating process in which metal coatings are deposited upon the surface of a cathodically polarized metal. Current densities used are roughly 2×10^{-2} Acm^{-2} and the thickness of the coatings is seldom more than about 10 μm. Current density in this context is regarded as the current used over the area under treatment. An example of an anodic dissolution operation is electropolishing, where the item to be polished is made the anode in an electrolytic cell; irregularities on its surface are dissolved preferentially so

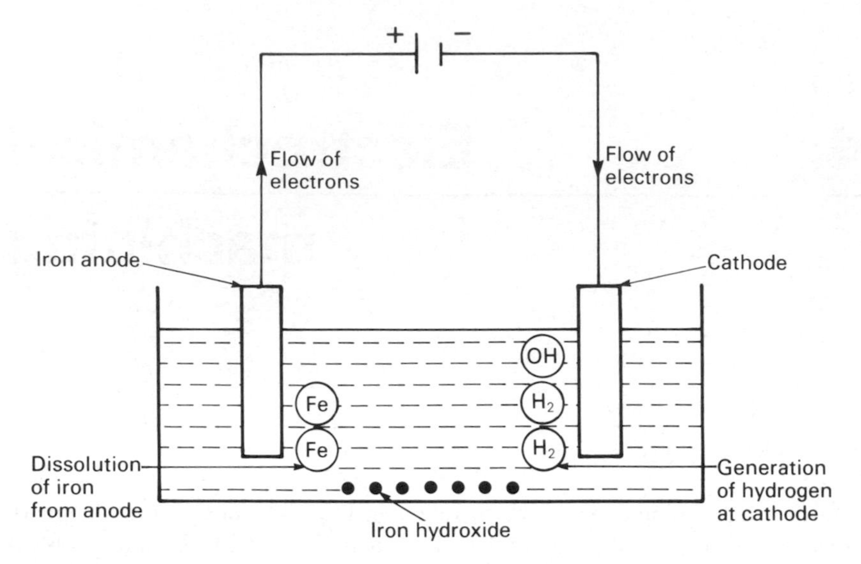

Fig. 4.1 Electrolysis of iron.

that, on their removal, the surface becomes flat and polished. A typical current density in this operation would be 10^{-1} Acm^{-2}, and polishing is usually achieved on the removal of irregularities as small as 10^{-2} μm. With both electroplating and electropolishing, the electrolyte is either in motion at low velocities or unstirred.

ECM is similar to electropolishing in that it also is an anodic dissolution process. But the rates of metal removal offered by the polishing process are considerably less than those needed in metal machining practice. To illustrate how ECM meets these requirements and, moreover, how it is used to shape metals, the electrolysis arising from iron in aqueous sodium chloride (Fig. 4.1) is now considered.

When a potential difference is applied across the electrodes, several possible reactions can occur at the anode and cathode. Certain reactions, however, are more likely to arise than others; this preference can be explained in terms of the *energy* that is available for each reaction. In the present example, the probable anodic reaction is dissolution of iron, e.g.

$$Fe \rightarrow Fe^{++} + 2e$$

At the cathode, the reaction is likely to be the generation of hydrogen gas and the production of hydroxyl ions:

$$2H_2O + 2e \rightarrow H_2 + 2OH^-$$

The outcome of these electrochemical reactions is that the metal ions combine with the hydroxyl ions to precipitate out as iron hydroxide, so that the net reaction is

$$Fe + 2H_2O \rightarrow Fe(OH)_2 + H_2$$

Note that the ferrous hydroxide may react further with water and oxygen to form ferric hydroxide:

$$4Fe(OH)_2 + 2H_2O + O_2 \rightarrow 4Fe(OH)_3$$

although it is stressed that this reaction, too, does not form part of the electrolysis.

With this metal-electrolyte combination, the electrolysis has involved the dissolution of iron from the anode, and the generation of hydrogen at the cathode; *no other action takes place at the electrodes.*

Certain observations relevant to ECM can be made at this stage:

1. Since the anode metal dissolves electrochemically, its rate of dissolution (or machining) depends, by Faraday's laws of electrolysis, only upon its atomic weight A, its valency z, the current I which is passed, and the time t for which the current passes. The dissolution rate is not influenced by the hardness or other characteristics of the metal.
2. Since only hydrogen gas is evolved at the cathode, the shape of that electrode remains unaltered during the electrolysis. This feature will be shown later to be most relevant in the use of ECM as a metal-shaping process.

4.3 DEVELOPMENT OF THE CHARACTERISTICS OF ECM

These two aspects can be developed further by use of Faraday's laws. If m is the mass of metal dissolved, and since $m = v\rho_a$, where v is the corresponding volume and ρ_a the density of the anode metal, the volumetric removal rate of anodic metal $\dot{v}$ is given by

$$\dot{v} = \frac{AI}{zF\rho_a} \tag{4.1}$$

Suppose that a machining operation has to be carried out on an iron workpiece at a typical rate, say $2.6 \times 10^{-8}\ m^3\ s^{-1}$. For this removal rate to be achieved by ECM, the current in the cell must be about 700 A (on substitution in equation (4.1) of the values $A/zF = 29 \times 10^{-8}\ kg\ C^{-1}$ and ρ_a 7860 kgm^{-3} for iron). Currents used in ECM are of this magnitude, and indeed they are often higher, by as much as an order of magnitude. The corresponding average current densities are typically 50 to 150 $\mathrm{A\,cm^{-2}}$.

The means by which high current densities are obtained can be understood from an examination of the other characteristics of an ECM cell, in particular the electrolyte conductivity and the inter-electrode gap width. These parameters are related to the current through Ohm's law, which states that the current I flowing in a conductor is directly proportional to the applied voltage V:

$$V = IR \tag{4.2}$$

In the simple expression (4.2), R is the resistance of the conductor. The experiments on electrolysis, described above, demonstrate that electrolytes are also conductors of electricity. Ohm's law also applies to this type of conductor, although the resistance of electrolytes may amount to hundreds of ohms.

Now the resistance R of a uniform conductor is directly proportional to its length h and inversely proportional to its cross-sectional area A. Thus

$$R = \frac{h\rho}{A} \tag{4.3}$$

where ρ is the constant of proportionality. If the conductor is a cube of side 10 mm, then $R = \rho$; ρ is termed the *specific resistance* or *resistivity* of the conductor. The reciprocal of the specific resistance is the specific conductivity, often denoted by the symbol κ.

If equations (4.2) and (4.3) are combined, the following relationships are derived between the average current density, current, surface area to be machined, applied potential difference, gap width and electrolyte conductivity, these quantities being denoted by the respective symbols J, I, A, V, h and κ_e:

$$J = \frac{I}{A} = \frac{\kappa_e V}{h} \tag{4.4}$$

It has been pointed out that in practice J is often about 50 $\mathrm{A\,cm^{-2}}$. To obtain a current density of this magnitude, a cell could be devised with high values for κ_e and V and low values for h. Even for strong electrolytes, however, κ_e is small. If the current is high, power requirements, amongst other considerations, restrict the use of high voltages, and, in practice, the voltage is usually about 10 to 20 V. If values of 0.2 $\Omega^{-1}\mathrm{cm^{-1}}$ and 10 V are taken for κ_e and V respectively, then for J to be 50 $\mathrm{A\,cm^{-2}}$ the gap h must be 0.4 mm. It will be shown later that a gap of this size is also necessary for accurate shaping of the anode. As dissolution of the anode proceeds, this gap is maintained by mechanical movement of one electrode, say the cathode, towards the other. To maintain the gap of 0.4 mm, a cathode feed-rate about 0.02 mms^{-1} would be needed, the values given above for the other process variables being retained.

The accumulation within the small machining gap of the metallic and gaseous products of the electrolysis is undesirable. If the growth were left uncontrolled, eventually a short circuit would occur between the two electrodes. To avoid this crisis, the electrolyte is pumped through the inter-electrode gap so that the products of the electrolysis are carried away. The forced movement of the electrolyte is essential also in diminishing the effects of electrical heating of the electrolyte, due to the passage of current, and of hydrogen gas, which respectively increase and decrease the effective conductivity. These matters will be discussed in greater detail later, but at this stage, the Joule heating effect provides a simple, convenient way of estimating a typical electrolyte velocity. Without forced agitation to control the increase in the electrolyte temperature, boiling will eventually occur in the gap. If all the heat caused by the passage of current remains in the electrolyte, the temperature increase δT, in a length δx of gap is, from Joule's and Ohm's laws,

$$\delta T = \frac{J^2 \delta x}{\kappa_e \rho_e c_e U} \tag{4.5}$$

where U is the electrolyte velocity, ρ_e the electrolyte density and c_e its specific heat.

If, for simplicity, the increase with temperature of the electrolyte conductivity is neglected, integration of equation (4.5) yields

$$U = \frac{J^2 L}{\kappa_e \rho_e c_e \Delta T} \tag{4.6}$$

where L is the electrode length and ΔT is the temperature difference of the electrolyte between points at inlet and outlet to the gap.

Consider the typical values, $J = 50\ \mathrm{Acm^{-2}}$, $L = 10^2$ mm, $\kappa_e = 0.2\ \Omega^{-1}\ \mathrm{cm^{-1}}$, $\rho_e = 1.1\ \mathrm{gcm^{-3}}$, $c_e = 4.18\ \mathrm{Jg^{-1}{}^\circ C^{-1}}$. Suppose, too, that ΔT must be kept to 75°C to avoid boiling at the exit point, the inlet temperature being, say, 25°C. From equation (4.6), the velocity to maintain this condition is calculated to be about 3.6 $\mathrm{ms^{-1}}$. Velocities of the electrolyte solution through the gap in ECM usually range from about 3 to 30 $\mathrm{ms^{-1}}$. The pressures required to achieve these velocities will now be calculated.

The electrolyte flow between the two electrodes in ECM is usually found to be turbulent, on the basis of the usual criterion that that condition occurs if

$$\mathrm{Re} > 2300 \tag{4.7}$$

where

$$\mathrm{Re} = \rho_e \frac{\bar{u} d_h}{\mu}$$

Re is the Reynolds number. Here, $\bar{u}$ is the mean velocity, d_h is the hydraulic mean diameter of the channel for the electrolyte flow, and μ is the absolute viscosity of the electrolyte, typically 1.19 cP.

Estimates of the viscous pressure drop down the flow channel of length L (electrode length) for turbulent flow may be obtained from

$$\frac{1}{2} \frac{f L \rho_e \bar{u}^2}{d_h} \tag{4.8}$$

where

$$f = \frac{0.3164}{\mathrm{Re}^{1/4}} \tag{4.9}$$

is the friction factor.

In a typical case, electrolyte of density $\rho_e = 1088\,\mathrm{kgm^{-3}}$, viscosity $\mu = 0.876\,\mathrm{cP}$ is pumped at a volume flow-rate of $0.98 \times 10^{-3}\,\mathrm{m^3 s^{-1}}$ between rectangular parallel electrodes of area $76.2 \times 38.1\,\mathrm{mm^2}$.

The mean velocity is calculated to be $27.96\,\mathrm{ms^{-1}}$. The Reynolds number is 64 000, with $d_h = 2h = 1.84\,\mathrm{mm}$. The flow rate is therefore turbulent. The friction factor, $f = 0.3164/\mathrm{Re}^{1/4}$, is estimated to be 0.0199.

The dynamic pressure, all of which is assumed to be lost at exit from the channel, is calculated to be

$$0.5 \rho_e \bar{u}^2 = 420\,\mathrm{kNm^{-2}}$$

The pressure needed to overcome friction effects is

$$\frac{1}{2} \frac{f \rho_e L u^2}{d_h} = 350\,\mathrm{kNm^{-2}}$$

Thus the total pressure needed to pump the electrolyte down the narrow channel formed by the cathode-tool and anode-workpiece is 770 kNm^{-2}. From the experimental work from which these data were taken, the actual pressure drop without ECM was 613 kNm^{-2}.

4.4 BASIC WORKING PRINCIPLES

ECM has been founded on the principles outlined in sections 4.2 and 4.3. As shown in Fig. 4.2, the workpiece and tool are made the anode and cathode, respectively, of an electrolytic cell, and a potential difference, usually fixed at about 10 V, is applied across them. A suitable electrolyte (e.g. aqueous NaCl

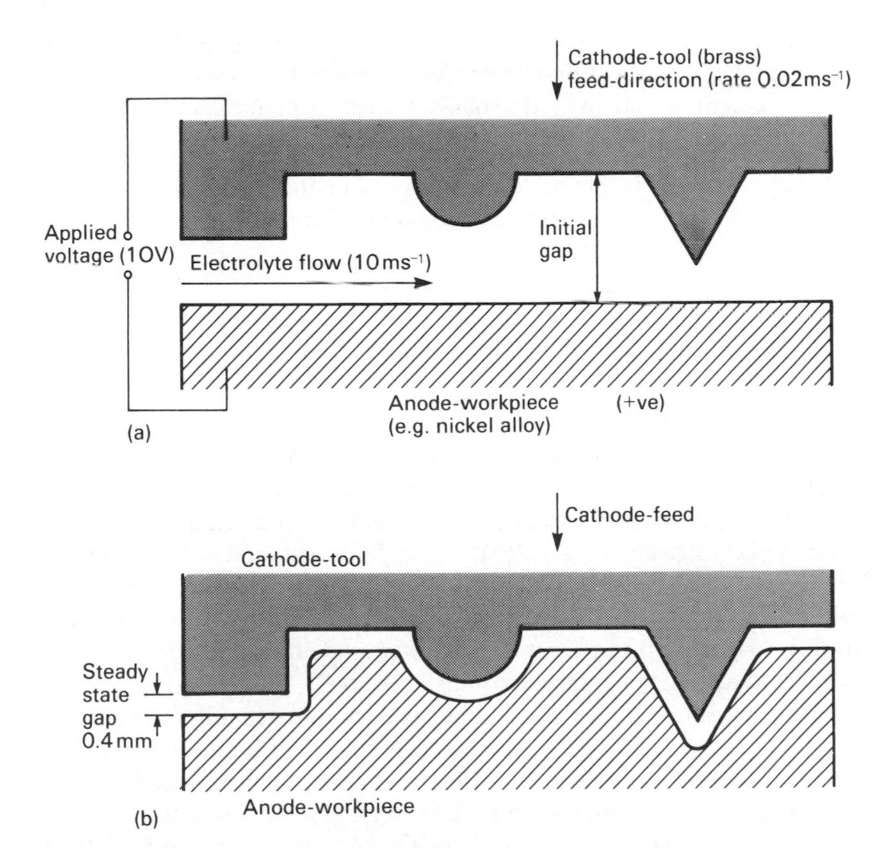

Fig. 4.2 Cathode-tool and anode-workpiece configurations for ECM.
(a) Initial.
(b) Final.

solution) is chosen so that the cathode shape remains unchanged during electrolysis. The electrolyte, whose conductivity is about $0.2\,\Omega^{-1}\,cm^{-1}$, is also pumped at a rate, roughly 3 to 30 ms^{-1} through the gap between the electrodes to remove the products of machining and to diminish unwanted effects, such as those that arise with cathodic gas generation and electrical heating. The rate at which metal is then removed from the anode is approximately in inverse proportion to the distance between the electrodes. As machining proceeds and, with the simultaneous movement of the cathode at a typical rate, say 0.02 mms^{-1}, towards the anode, the gap width along the electrode length will gradually tend to a steady-state value. Under these conditions, a shape, roughly complementary to that of the cathode, will be reproduced on the anode. A typical gap width then should be about 0.4 mm and the average current density should be of the order of 50 to 150 Acm^{-2}. Moreover, if a complicated shape is to be formed on a workpiece of a hard material, the complementary shape can first be produced on a cathode of softer metal, and the latter electrode is then used to machine electrochemically the workpiece. In short, the main advantages of ECM are:

1. the rate of metal machining does not depend on the hardness of the metal;
2. complicated shapes can be machined on hard metals;
3. there is no tool wear.

4.5 INDUSTRIAL ELECTROCHEMICAL MACHINE

Figure 4.3 shows industrial electrochemical machines, the working principles of which are based on the principles given above. A small 500 A machine is depicted in Fig. 4.3(a). It is suitable for the machining of light items, for instance for deburring and hole drilling (section 4.10). Such a machine needs the components given in Fig. 4.3(b). Particular attention has to be paid to the stability of the electrochemical machine tool frame, and to the machining table which should also be stable and firm. The electrolyte has to be filtered carefully to remove the products of machining and often has to be heated in its reservoir to a fixed temperature, for instance 30°C. This procedure is used to provide constant operating conditions, as during machining the electrolyte heats up, due to the passage of current. If precautions are not taken the high electrolyte temperature can cause too great changes in its specific conductivity, with subsequent undesirable effects on machining accuracy.

At the other end of the scale from Fig. 4.3(a), a substantial 10 000 A electrochemical machining installation is given in Fig. 4.3(c), although even this machine still retains the basic components given in Fig. 4.3(b). This large machine is used for a range of aircraft engine components (section 4.10).

4.6 RATES OF MACHINING

Faraday's laws can be employed to calculate the rates at which metals can be electrochemically machined. His laws are embodied in the simple expression

$$m = \frac{A}{z}\frac{It}{F} \tag{4.10}$$

where m is the mass of metal electrochemically machined by a current, I(A) passed for a time, t(s). A is the atomic weight of the dissolving ions, z is their valency, and F is Faraday's constant (= 96 500 C). The quantity A/zF is called the electrochemical equivalent of the anode-metal.

Table 4.1 shows the metal machining rates that can be obtained when a current of 1000 A is used in ECM. Note that the metal removal rates are given in terms of volumetric machining rates as well as mass removal rates. The former is often more useful in practice. Theoretical volumetric rates $\dot{v}$ can be calculated from equation (4.1) given above.

Table 4.1 assumes that all the current is used in ECM to remove metal. Unfortunately that is not always the case. Some metals are more likely to machine at the Faraday rates of dissolution than others. Many factors, other than current, influence the rate of machining. They involve electrolyte type, the rate of electrolyte flow and other process conditions.

For example, nickel will machine at 100% current efficiency (defined as the percentage ratio of the experimental to theoretical rates of metal removal) at low current densities (e.g. 25 Acm^{-2}). If the current density is increased (say to 250 Acm^{-2}) the efficiency is found to be reduced to typically 85 to 90%, by the onset of other reactions at the anode, such as oxygen gas evolution, which become increasingly preferred as the current density is increased.

If the ECM of titanium is attempted in sodium chloride electrolyte, very low current efficiencies, of about 10 to 20%, are usually obtained. When this solution is replaced by some mixture of fluoride-based electrolytes, higher efficiencies can be obtained, albeit by the additional measure of higher voltages (roughly 60 V) which are also needed to help break down the tenacious oxide film that forms on the surface of this metal. Ironically this film accounts for the corrosion resistance of titanium, which together with its toughness and lightness, makes this metal so useful in the aircraft engine industry.

If the rates of electrolyte flow are kept too low, the current efficiency of even the most easily electrochemically machined metal is reduced, since with insufficient flow the products of machining cannot be so readily flushed from the machining gap. The accumulation of debris within the gap impedes the further dissolution of metal, and the build-up of cathodically generated gas

(a)

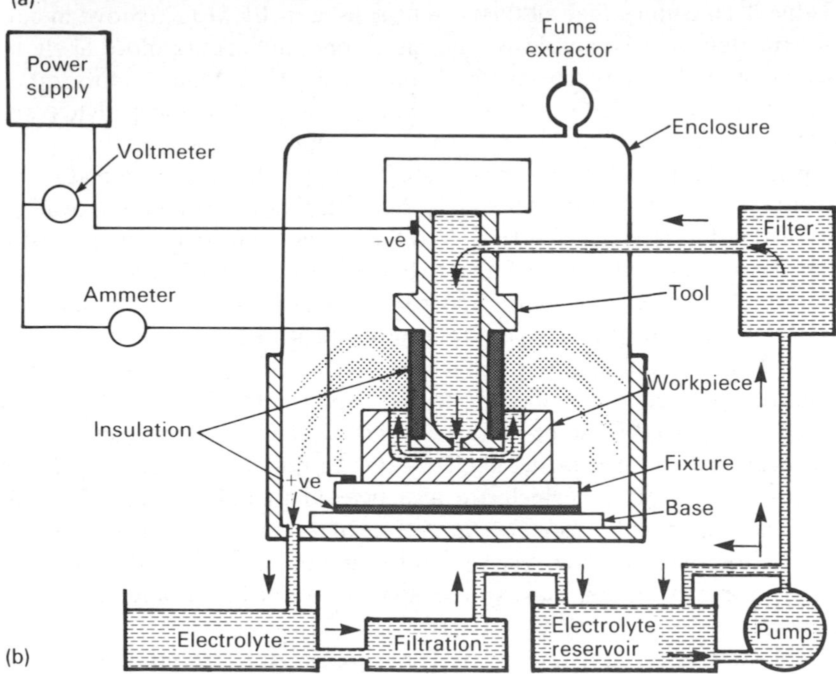

(b)

Fig. 4.3 Industrial electrochemical machine.
(a) small electrochemical machine (500 A unit).
(b) components of industrial machine.
(c) large industrial electrochemical machine (10 000 A).
(By courtesy of Rolls-Royce, plc.)

(c)

can lead to short-circuiting between the tool and workpiece, causing termination of machining and damage to both electrodes if the feed mechanism is not stopped. Indeed the inclusion of proper flow channels to provide sufficient flow of electrolyte so that machining can be efficiently maintained remains a major exercise in ECM practice. When complex shapes have to be produced the design of tooling incorporating the right kind of flow ports becomes a considerable problem, requiring skill and process experience by the design engineers.

Although Table 4.1 provides data on machining rates for pure metals, various expressions have been derived from which the corresponding rates for alloys can be calculated. All these procedures are based on calculating an effective value for the chemical equivalent of the alloy. Thus for Nimonic 75, a typical nickel-alloy used in the aircraft industry, a chemical equivalent of 25.1 may be derived from the expression

chemical equivalent of alloy =

$$\left\{100/\left[\frac{X_A}{A_A/z_A}+\frac{X_B}{A_B/z_B}+\dots\right]\right\} \tag{4.11}$$

Table 4.1 Theoretical removal rates for a current of 1000 A

Metal	*Atomic weight*	*Valency*	*Density ($kg\ m^{-3} \times 10^3$)*	*Removal rate* ($kg\ s^{-1} \times 10^{-6}$)	10^{-6} (m^3/s)
Aluminium	26.97	3	2670	95	0.035
Beryllium	9.0	2	1850	50	0.025
Chromium	51.99	2	7190	250	0.038
		3		200	0.025
		6		90	0.013
Cobalt	58.93	2	8850	305	0.035
		3		205	0.023
Niobium	92.91	3	9570	320	0.034
(Columbium)		4		240	0.025
		5		195	0.020
Copper	63.57	1	8960	660	0.074
		2		330	0.037
Iron	55.85	2	7860	290	0.037
		3		195	0.025
Magnesium	24.31	2	1740	125	0.072
Manganese	54.94	2	7430	285	0.038
		4		140	0.019
		6		95	0.013
		7		80	0.011
Molybdenum	95.94	3	10 220	330	0.032
		4		250	0.024
		6		165	0.016
Nickel	58.71	2	8900	305	0.034
		3		205	0.023
Silicon	28.09	4	2330	75	0.031
Tin	118.69	2	7300	615	0.084
		4		305	0.042
Titanium	47.9	3	4510	165	0.037
		4		125	0.028
Tungsten	183.85	6	1930	315	0.016
		8		240	0.012
Uranium	238.03	4	1910	620	0.032
		6		410	0.022
Zinc	65.37	2	7130	340	0.048

Note: In ECM currents from 250 to 10 000 A are common.

where A, B, . . . denote the elements in the alloy of percentage amounts X. (The Nimonic alloy is given to have the percentage of weight constituents 72.5 Ni, 19.5 Cr, 5.0 Fe, 0.4 Ti, 1.0 Si, 1.0 Mn, 0.5 Cu.)

4.7 SURFACE FINISH IN ECM

As well as influencing the rate of metal removal, the electrolytes also affect the quality of surface finish obtained in ECM, although other process conditions also have an effect, as will be discussed below.

Depending on the metal being machined, some electrolytes leave an etched finish, caused by the non-specular reflection of light from crystal faces electrochemically dissolved at different rates. Sodium chloride electrolyte tends to produce a kind of etched, matt finish with steels and nickel alloys: a typical surface roughness would be about 1 μm Ra.

In many applications, a polished finish is desirable on machined components. The production of an electrochemically polished surface is usually associated with the random removal of atoms from the anode-workpiece, the surface of which has become covered with an oxide film. They are determined by the particular metal-electrolyte combination being used. The mechanisms controlling high-current density electropolishing in ECM are still not completely understood. For example with nickel-based alloys, the formation of a nickel oxide film seems to be necessary; a polished surface, with roughness of 0.2 μm Ra has been claimed for a Nimonic (nickel alloy) machined in saturated sodium chloride solution. Surface finishes as fine as 0.1 μm Ra have been reported when nickel-chromium steels have been machined in sodium chlorate solution: the formation of an oxide film on the metal surface has been considered to be the key to these conditions of polishing.

But sometimes the formation of oxide films on the metal surface hinders efficient ECM, and leads to poor surface finish. For example, the ECM of titanium is rendered difficult in chloride and nitrate electrolytes because the oxide film formed is so passive. Even when higher voltages are applied, e.g. about 50 V, to break the oxide film its disruption is so non-uniform that deep grain boundary attack of the metal surface occurs.

Occasionally, metals that have undergone ECM are found to have a pitted surface, the remaining area being polished or matt. Pitting normally stems from gas evolution at the anode-electrode; the gas bubbles rupture the oxide film causing localized pitting.

Process variables also play a significant part in the determination of the surface finish. For example, as the current density is raised, generally, the smoother the finish on the workpiece surface becomes. For instance, tests with nickel machined in HCl solution have shown that the surface finish improves from an etched to a polished appearance when the current density is increased from about 8 to 19 Acm^{-2}, the flow velocity being held constant. A similar effect is achieved when the electrolyte velocity is increased.

4.8 ACCURACY AND DIMENSIONAL CONTROL

ECM is not a highly accurate process. The distribution of the electric current lines leads to rounding of edges. Thus very sharp corners cannot be produced by ECM; tolerances of about 0.127 mm are generally held to be typical. Accuracies to 0.013 mm have been claimed under special circumstances. Recent reports on micro-ECM have revealed that accuracy of ECM can be improved by special shielding and masking in order to direct the current flow only to required areas (Landolt, 1987).

Electrolyte selection plays an important role. Sodium chloride, for example, yields much less accurate components than nitrate, the latter electrolyte having far better dimensional control due to its current efficiency/current density characteristics.

As indicated in Fig. 4.4, with sodium nitrate electrolyte, the current efficiency is greatest at the highest current densities. In hole drilling these high current densities occur between the leading edge of the drilling tool and the workpiece. In the side gap there is no direct movement between the tool and workpiece surface, so that gap widens and the current densities are lower. The

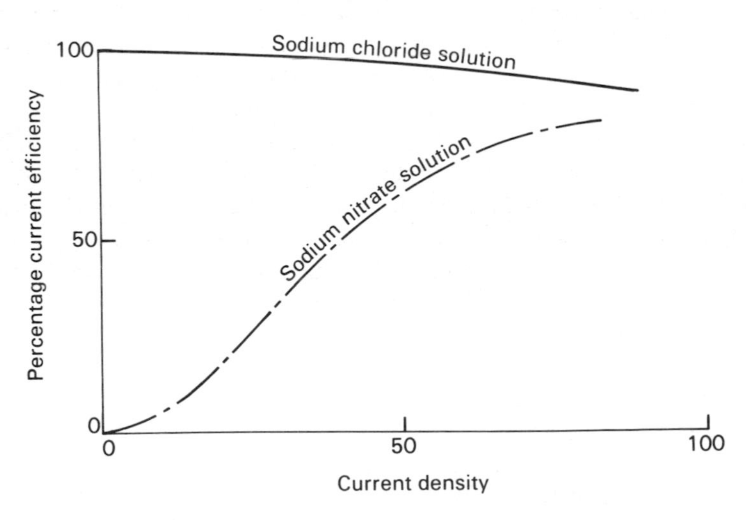

Fig. 4.4 Schematic illustration of effects of different electrolytes on the current efficiency/current density relationship.

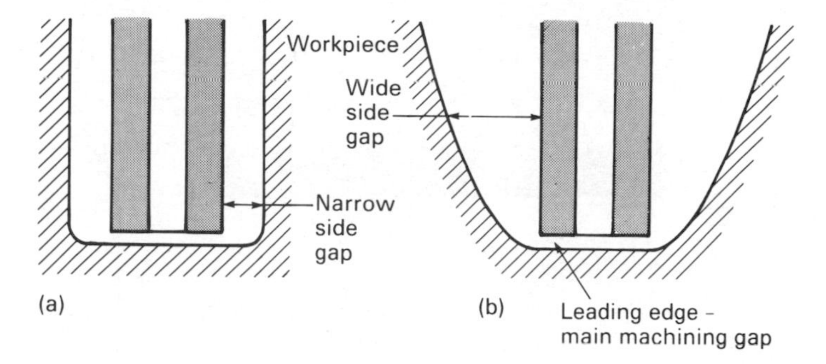

Fig. 4.5 Schematic representation of effects of different electrolytes on overcut in drilling.
(a) Sodium nitrate.
(b) Sodium chloride.

current efficiencies are consequently lower there; much less metal than that predicted from Faraday's law is removed. Thus, as shown in Fig. 4.5, the overcut in the side gap is reduced with this type of electrolyte. If another electrolyte such as sodium chloride solution were used instead, then the overcut could be much greater. Its current efficiency remains steady at almost 100% for a wide range of current densities. Thus even in the side gap, metal removal proceeds at a rate which is mainly determined by only the current density, in accordance with Faraday's law. A wider overcut then ensues.

Other electrolytes that are used include mild (about 5%) hydrochloric acid solution, which is particularly useful in fine-hole drilling, since this acid electrolyte dissolves the metal hydroxides as they are produced; it resembles NaCl electrolyte in that its current efficiency is about 100%. Sodium chlorate solution has also been investigated. Industrialists have been reluctant to employ it as an ECM electrolyte, because of its ready combustibility. None-theless this electrolyte is claimed to give even better throwing power and closer dimensional control than sodium nitrate solution.

4.9 THEORY OF SHAPING IN ECM

Most metal-shaping operations in ECM utilize the same inherent feature of the process whereby one electrode, the cathode-tool say, is driven towards the other at a constant rate, a fixed voltage being applied between them. Under these process conditions, the gap width between the tool and the workpiece will become constant, the rate of forward movement of the tool

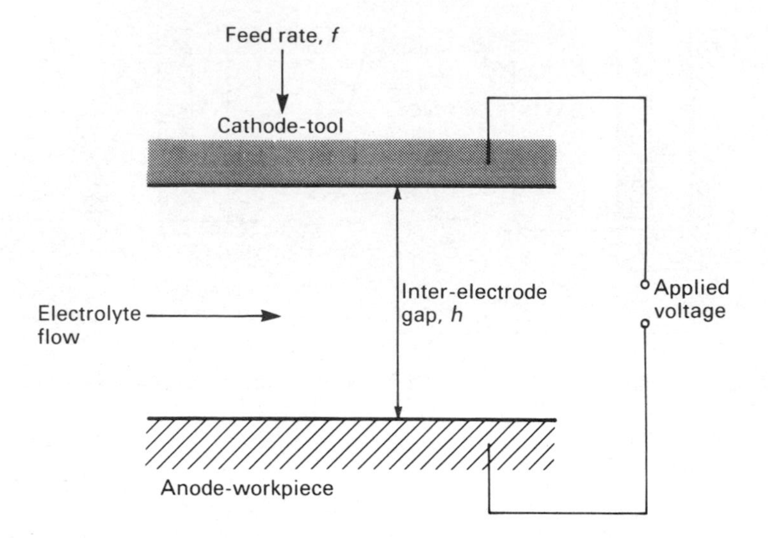

Fig. 4.6 Plane parallel electrodes in ECM.

being matched by the rate of recession of the workpiece surface, due to electrochemical dissolution.

Some useful expressions can be derived for the variation of the inter-electrode gap width. As indicated in Fig. 4.6, consider a set of plane-parallel electrodes, with a constant voltage V applied across them, and with the cathode-tool driven mechanically towards the anode-workpiece at a constant rate f. In this analysis, the electrolyte flow is not expected to have any significant effect on the specific conductivity of the electrolyte κ_e which is assumed to stay constant throughout the ECM operation. Also, all the current that is passed is taken to be used to remove metal from the anode, i.e. no other reactions occur there.

Under these conditions, from Faraday's law, the rate of change of gap width h relative to the tool surface is

$$\frac{dh}{dt} = \frac{AJ}{ZF\rho_a} - f \tag{4.12}$$

where A, Z are the atomic weight and valency respectively of the dissolving ions, F is Faraday's constant, ρ_a is the density of the anode-workpiece metal, and J is the current density.

From Ohm's law, the current density J is given by

$$J = (\kappa_e V/h) \tag{4.13}$$

where h is the gap-width between the electrodes.

On substitution of equation (4.13) into (4.12) we have

$$\frac{dh}{dt} = \frac{A\kappa_e V}{ZF\rho_a h} - f \tag{4.14}$$

Three practical cases are of interest in considering solutions to equation (4.14):

(a) Feed-rate $f = 0$, i.e. no tool movement

Equation (4.14) then has the solution for gap $h(t)$ at time (t).

$$h^2(t) = h^2(0) + \frac{2A\kappa_e Vt}{ZF\rho_a} \tag{4.15}$$

where $h(0)$ is the initial machining gap. That is, the gap width increases indefinitely with the square root of machining time, t, see Fig. 4.7(a). This condition is often used in deburring by ECM when surface irregularities are removed from components in a few seconds, without the need for mechanical movement of the electrode.

(b) Constant feed-rate f

The tool is removed mechanically at a fixed rate towards the workpiece. Equation (4.14) then has the solution

$$t = \frac{1}{f}\left\{h(0) - h(t) + h_e \ln_e \frac{h(0) - h_e}{h(t) - h_e}\right\} \tag{4.16}$$

Note that the gap width tends to a steady state value, h_e given by

$$h_e = \left(\frac{A\kappa_e V}{ZF\rho_a f}\right) \tag{4.17}$$

This inherent feature of ECM whereby an equilibrium gap width is obtained, is used widely in ECM for reproducing the shape of the cathode-tool on the workpiece.

This solution to equation (4.16) is drawn schematically in Fig. 4.7(b).

(c) Short-circuit conditions

If process conditions arise such as too high a feed-rate, the equilibrium gap then obtained may be so small that contact between the two electrodes

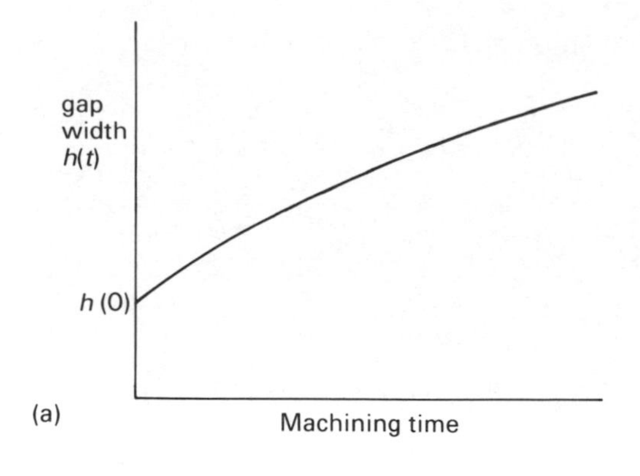

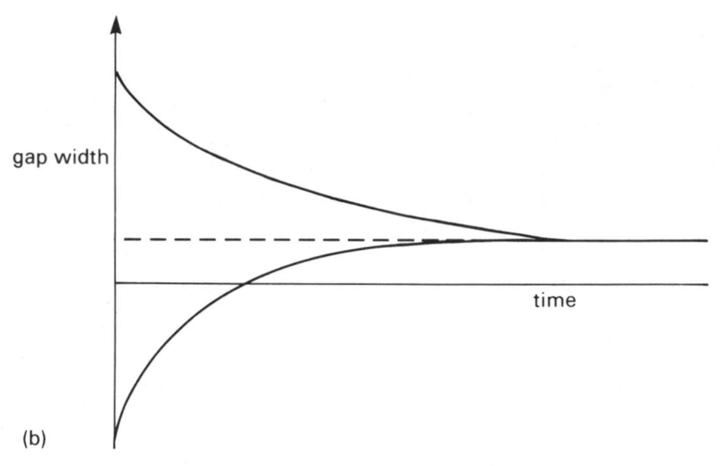

Fig. 4.7 Variation of gap width with time of machining.
(a) Increase in gap width with time (zero electrode feed rate).
(b) Attainment of equilibrium gap-width (constant feed rate).

ensues. This condition would cause a short-circuit between the electrodes and hence premature termination of machining.

The equilibrium gap is applied widely in the shaping process. Studies of ECM shaping are usually concerned with three distinct problems:

1. By far the primary problem is design of a cathode-tool shape needed to produce a required profile geometry of the anode-workpiece. An example is shown in Fig. 4.8.
2. For a given cathode-tool shape, prediction of the resultant anode-workpiece geometry. A common example here is hole-drilling by ECM. A tube-shaped cathode-drill yields a workpiece profile shown in Fig. 4.9.

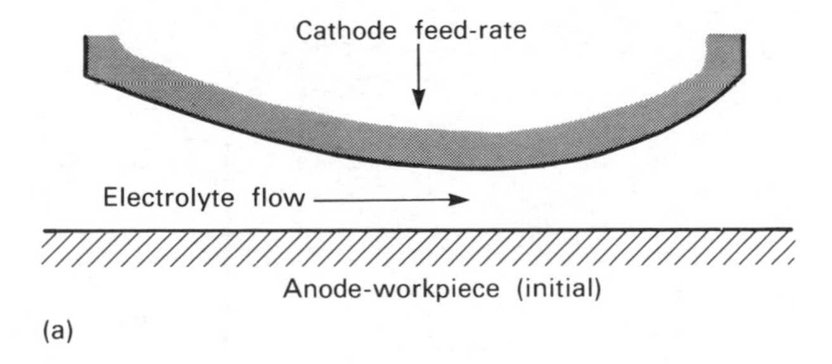

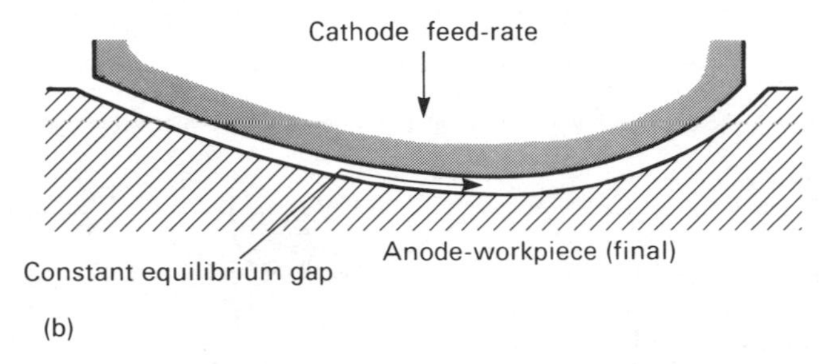

Fig. 4.8 Design of cathode-tool shape. Initially this shape is made complementary to that needed on the workpiece; design has to take account of effects of electrolyte flow which can cause non-uniform machining of profile.
(a) Initial.
(b) Final.

3. Specification of the shape of the anode workpiece, as machining proceeds. This is most readily predicted for the smoothing of surfaces, as illustrated in Fig. 4.10.

Computational and theoretical analyses of the problems are based on three basic equations, including Laplace's equation

$$\nabla^2 \varphi = 0 \tag{4.18}$$

the solution of which gives the potential φ at any point in the electrolyte, particularly at the electrode surfaces.

From the potential so found, the current density J can be obtained from Ohm's law, written in the form

$$J = -\kappa_e \nabla \varphi \tag{4.19}$$

where κ_e is the electrolyte conductivity.

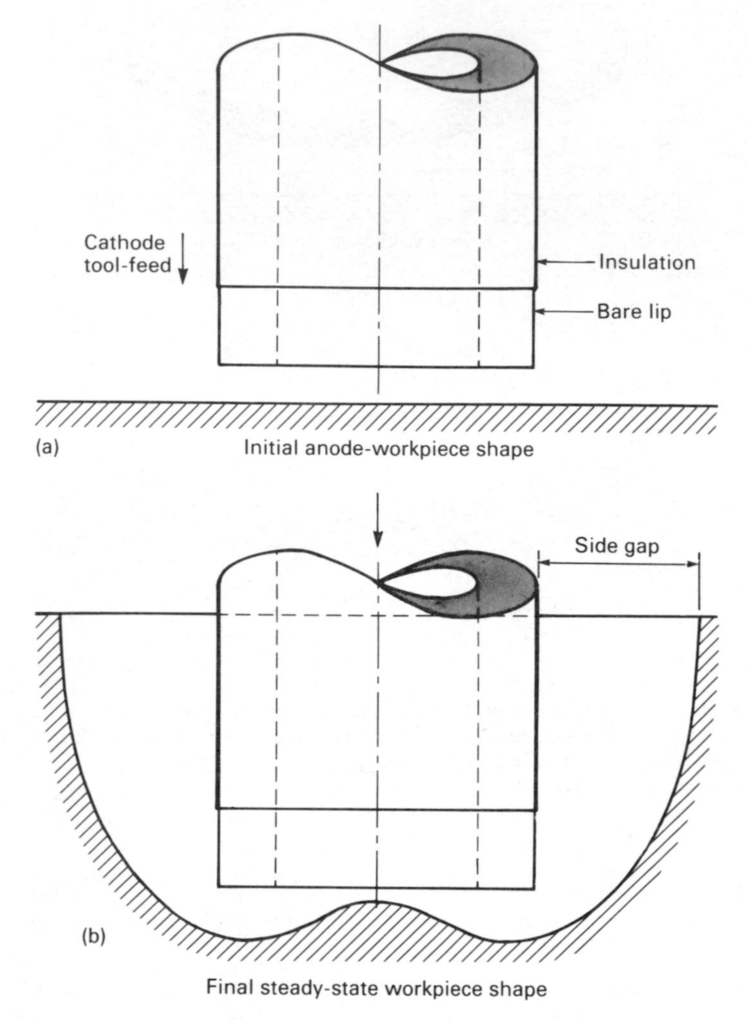

Fig. 4.9 Prediction of resultant anode profile for given cathode-tool shape.
(a) Initial.
(b) Final condition.

Finally, Faraday's law

$$\dot{r}_a = \frac{AJ}{ZF\rho_a} \tag{4.20}$$

is then used to give the anode recession rate $\dot{r}_a$, A being the atomic weight, Z the valency, ρ_a the density of the anode metal and F Faraday's constant.

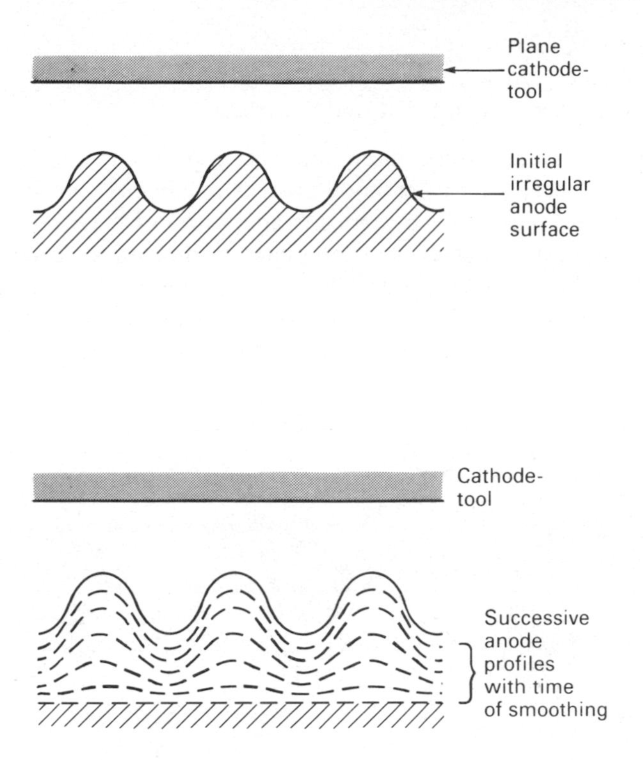

Fig. 4.10 Specification of anode profile with time of machining.

If the electrode potentials are equipotentials, the boundary conditions are

$$\varphi = 0 \text{ at the cathode-tool}$$

$$\varphi = V \text{ at the anode-workpiece}$$

where V is the applied potential difference. Overpotentials at these electrodes alter those boundary conditions by appropriate amounts – see McGeough (1974).

Solutions to these equations have been attempted in several ways, including fully analytic, graphical-analogue and computational methods. The last mentioned are undoubtedly the most practicable. However the complexity of the shaping problem in ECM still renders difficult the application of these theoretical methods to practical problems. As a result, empirical design of tool-shapes to effect a required workpiece configuration is still widespread.

4.10 APPLICATIONS

4.10.1 Deburring (smoothing of rough surfaces)

The simplest and a very common application is deburring. An example is given in Fig. 4.11, where a cathode-tool complementary in shape is placed opposite a workpiece, which carries irregularities of height 0.25 mm on its surface. The current densities at the peaks of the surface irregularities are higher than those elsewhere. The former are therefore removed preferentially, and the workpiece becomes smoothed. In this application, a current of 250 A at 12 V was applied to remove 0.08 mm of metal from the irregularity in

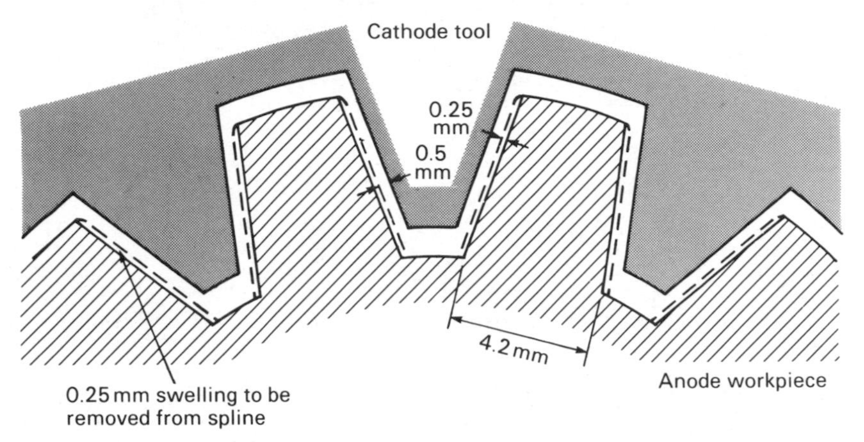

Fig. 4.11 Surface irregularities on external spline of case-hardened steel shift hub sleeve. Configuration of cathode tool and external spline carrying surface irregularity. (After McGeough, 1974.)

about 3 s. A 30% (w/w) $NaNO_3$ electrolyte at 30°C was used. Its good dimensional control restricted ECM mainly to the region of the irregularities.

Electrochemical deburring is a fast process: typical process times are 5 to 30 s for smoothing the surfaces of manufactured components. Owing to its speed and simplicity of operation, electrochemical deburring can often be performed with a fixed, stationary cathode-tool. The process is used in many applications, and is particularly attractive for the deburring of the intersectional region of cross-drilled holes, see Fig. 4.12.

4.10.2 Hole drilling

Hole drilling is another popular way of using ECM. As indicated in Fig. 4.13 a tubular electrode is used as the cathode-tool. Electrolyte is pumped down the

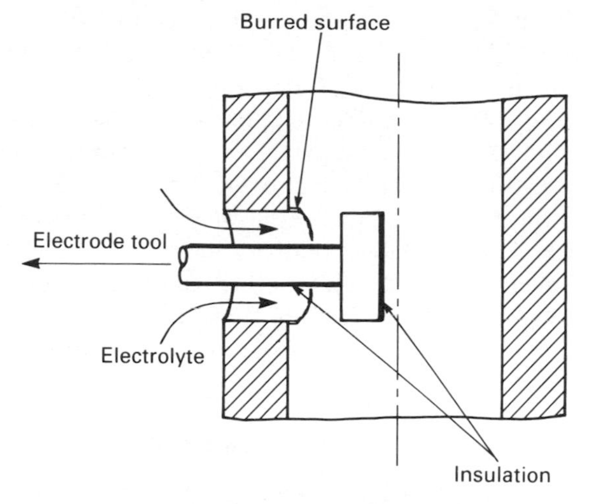

Fig. 4.12 Electrochemical deburring of cross-drilled hole.

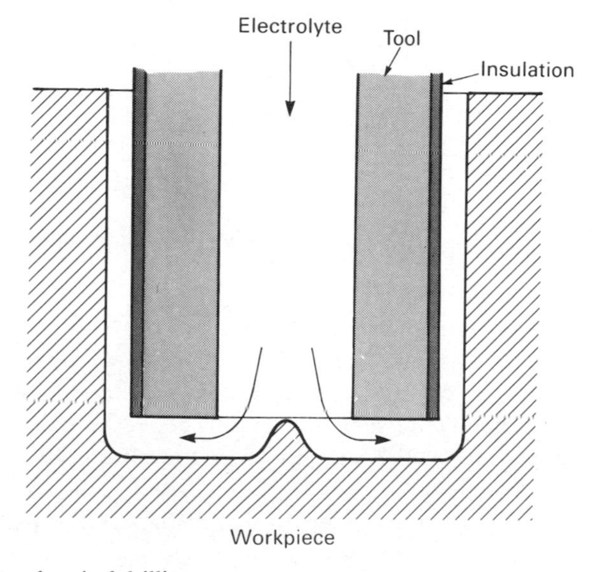

Fig. 4.13 Electrochemical drilling.

central bore of the tool, and out through the side-gap formed between the wall of the tool and the hole electrolytically dissolved in the workpiece.

The main machining action is carried out in the inter-electrode gap formed between the leading edge of the drill-tool and the base of the hole in the workpiece. ECM also proceeds laterally between the side walls of the tool and component. Since the lateral gap width is larger than that at the leading edge of the tool the side ECM rate is lower. The overall effect of the side ECM is to increase the diameter of the hole that is produced. The local difference between (*a*) the radial length between the side wall of the workpiece and the central axis of the cathode-tool, and (*b*) the external radius of the cathode, is known as the 'overcut'. The amount of overcut can be reduced by several means. A common method is by reversal of the direction of flow of the electrolyte, which is pumped down between the outer wall of the cathode-tool, across the main machining gap and then upwards through the central bore of the drilling-tool. This procedure removes the gaseous products of electrolysis from the machining zone without their reaching the side-gap. The

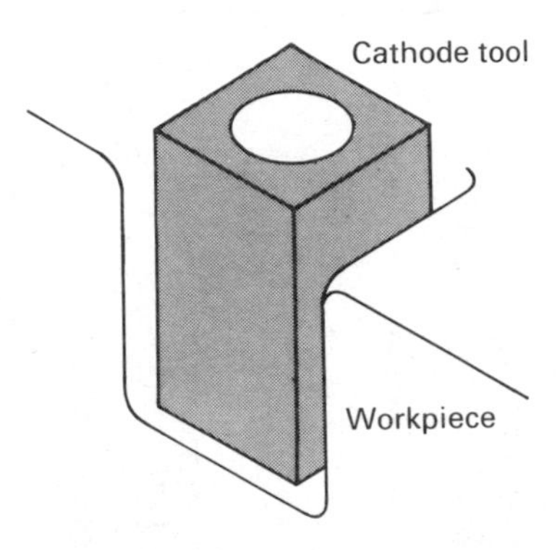

Fig. 4.14 Cathode-drill (square section) gives corresponding square hole.

overcut can also be reduced by electrical insulation of the external walls of the tool which thereby stops side current flows. A third way of reducing overcut is by the choice of an electrolyte like sodium nitrate, which permits high, efficient rates of metal removal in the leading edge where the current density is high, and lower rates in the side gap where the current density is lower (see Fig. 4.5).

A wide range in hole-sizes can be drilled: diameters as small as 0.05 mm to as large as 20 mm have been reported.

Drilling by ECM is not restricted to round holes. Since the shape of the workpiece is determined by that of the tool-electrode, as previously discussed, a cathode-drill with any cross-section, will produce a corresponding shape on the workpiece (Fig. 4.14).

The significant differences between drilling and smoothing are noteworthy. With the former, forward mechanical movement of one of the electrodes, say the tool, towards the other is usually necessary in order to maintain a constant equilibrium gap width in the main machining zone between the leading face of the drill and the workpiece. In smoothing, mechanical drive of the tool can often be avoided. A typical feed-rate in ECM-drilling is 1–5 mm min^{-1}.

4.10.3 Full-form shaping

Unlike the previous two techniques, the third method in ECM, 'full-form shaping', utilizes a constant gap across the entire workpiece, and a constant feed-rate in order to produce the type of shape illustrated in Fig. 4.15. Here the current density remains high across the entire face of the workpiece.

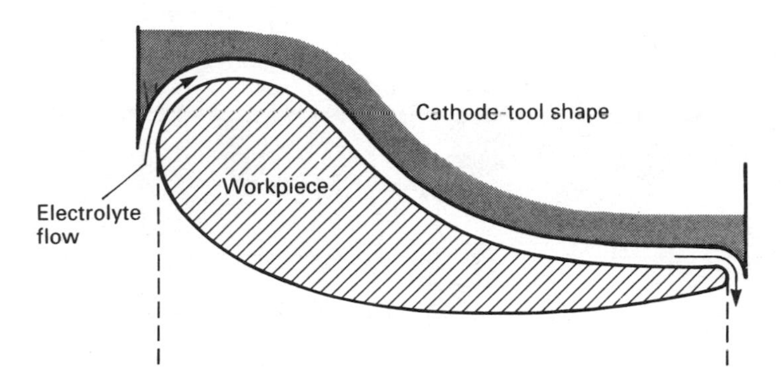

Fig. 4.15 Design of cathode tool shape to obtain profile of turbine blade. (Design has to take into account effects of electrolyte flow on conductivity of solution which can cause non-uniform machining of profile.)

Full-form shaping is well known for the production of compressor and turbine blades and in this procedure, current densities as high as 100 A cm^{-2} are used.

Electrolyte flow plays an even more influential role in full-form shaping than in drilling and smoothing. The entire larger cross-sectional area of the workpiece has to be supplied by the electrolyte as it flows between the electrodes. The larger areas of electrodes involved mean that comparatively higher pumping pressures and volumetric flow rates are needed.

A fully automated ECM system for the manufacture of thin, accurately profiled high-performance compressor blades has been developed. As indicated in Fig. 4.16, the blade root is formed from cut titanium or nickel alloy bar by broaching, and then the blade aerofoil is shaped by ECM. Another advanced machining method, Electro-Discharge Machining (EDM – see Chapter 6), is employed to cut the formed blade to length.

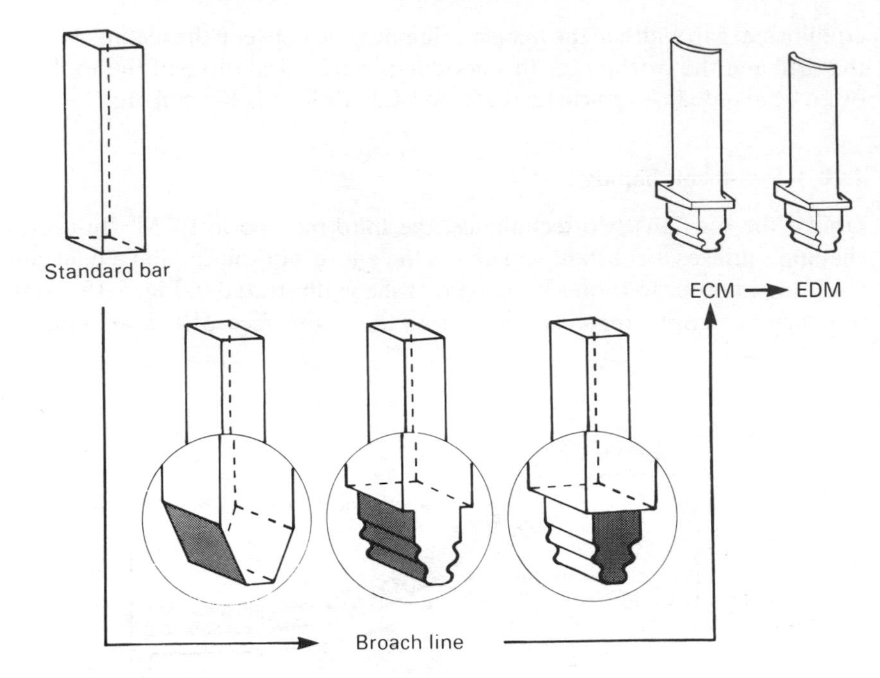

Fig. 4.16 Stages in the production of a compressor blade (by kind permission of Rolls-Royce, plc, Bristol).

Throughout the process the component is transferred to each station by robot (Fig. 4.17). The removal of sharp edges and quality control are automatic. This ECM system has made possible the economic production of small complicated compressor blades which has not always been possible by conventional means. This procedure, which is also known as the 360° ECM system, permits each finished blade to be produced from the raw bar in a machining cycle-time of 4 minutes.

4.10.4 Wire-machining

Clifton *et al.* (1987) have described a system in which a wire-electrode is used in ECM to remove boat-shaped samples for metallurgical examination.

4.11 SPECIAL ECM APPLICATIONS

Examples of advanced or modified versions of the basic ECM process may be usefully described here.

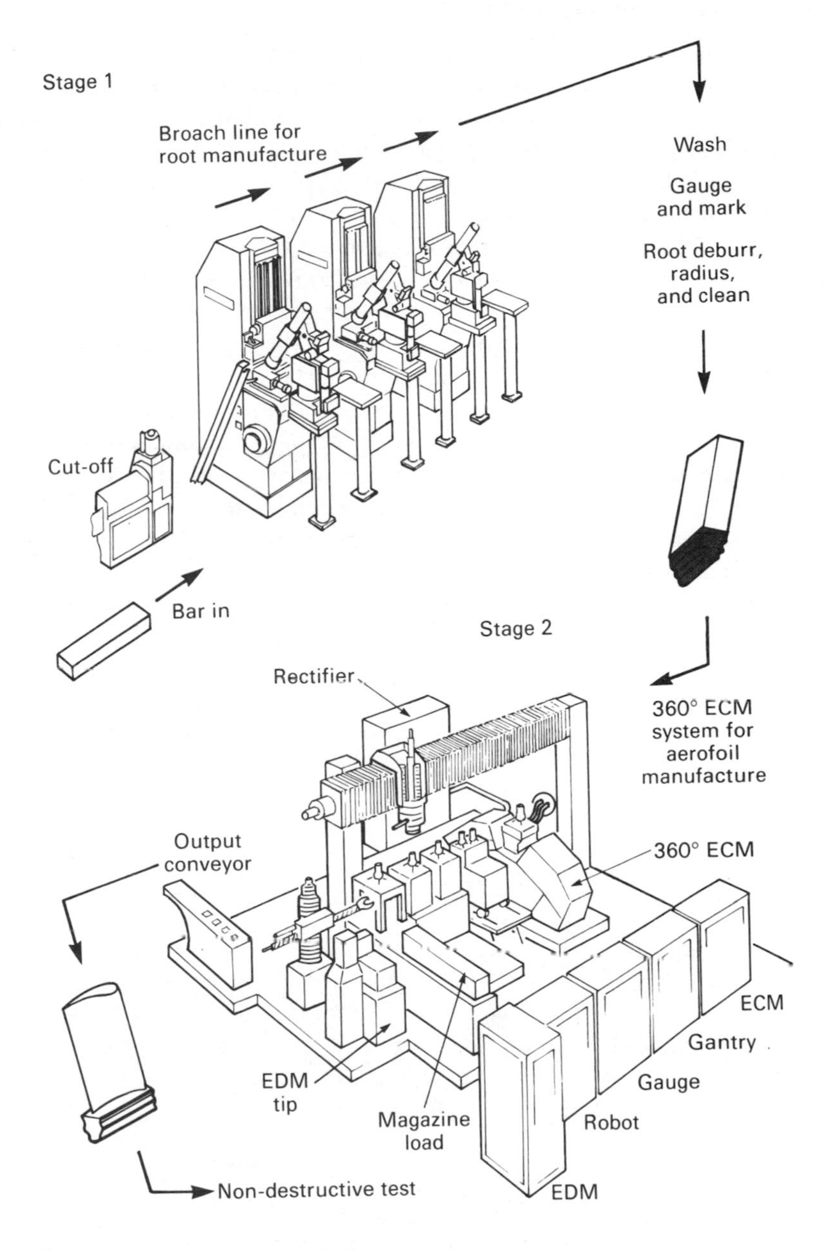

Fig. 4.17 Automatic ECM system for manufacturing small compressor blades (by kind permission of Rolls-Royce, plc, Bristol).

4.11.1 Electrochemical grinding (ECG)

The basic apparatus used in this aspect of ECM is shown in Fig. 4.18. The main feature is the use of a grinding wheel in which an insulating abrasive, such as diamond particles, is set in a conducting bonding material. This wheel becomes the cathode-tool. The non-conducting particles act as a spacer between the wheel and workpiece, providing a constant inter-electrode gap, through which electrolyte is flushed.

When a voltage of about 4 to 8 V is applied between the wheel and the workpiece, current densities of about 120 to 240 $A cm^{-2}$ are created, removing metal mainly by ECM, although mechanical action of the non-conducting particles accounts for an additional 5 to 10% of the total metal removal. The rate of machining is typically 1600 $mm^3 min^{-1}$. The surface finish produced by ECG varies from 0.2 μm to 0.3 μm, depending on the metal being ground.

Accuracies achieved by ECG are usually about 0.125 mm, although some claims have been made for accuracies an order of magnitude better. A drawback of ECG is the loss of accuracy when inside corners are ground; because of the electric field effects, radii better than 0.25 to 0.375 mm can seldom be achieved.

A wide application of electrochemical grinding is the production of tungsten carbide cutting tools. ECG is also useful in the grinding of fragile parts such as hypodermic needles and thin-wall tubes.

A recent application of the technique has arisen in the offshore industry, for the removal of fatigue cracks from underwater steel structures. Seawater

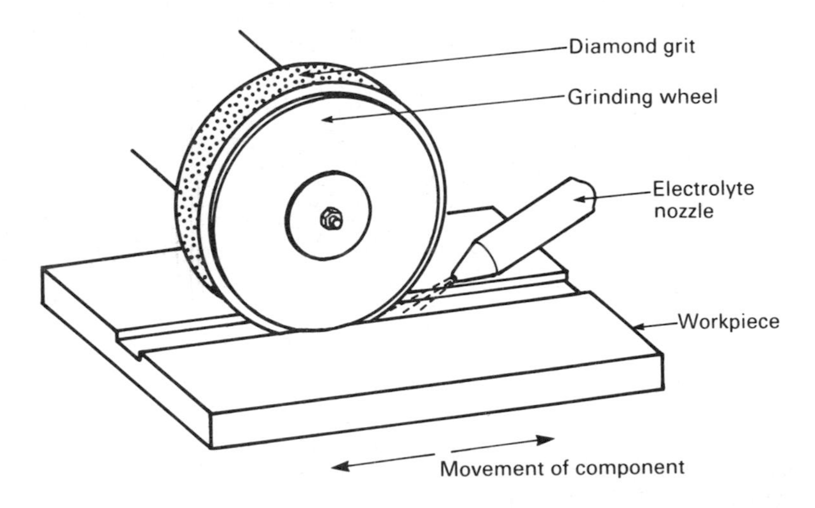

Fig. 4.18 Electrochemical grinding.

itself is an electrolyte, being composed mainly of sodium chloride solution of approximate salinity 3.5%. Although its specific conductivity is about one-fifth of that of electrolytes normally used in ECM, it is a suitable vehicle for ECG, and is used in the North Sea. The diamond particles embedded in the grinding tool are used to remove non-conducting materials, such as organic sea growth on the surface of the steel, before the ECG action properly starts. Holes about 25 mm in diameter, in steel 12 to 25 mm thick, have been produced by ECG at the ends of fatigue cracks to stop further development of the crack and to enable the removal of specimens for metallurgical inspection.

4.11.2 Shaped-tube electrolyte machining (STEM)

The drilling of cooling holes of large depth-to-diameter ratio in high-pressure turbine blades has been accomplished by 'shaped-tube electrolytic machining'. The technique is usually called 'STEM drilling'.

The turbine blades are cooled by passing cool air through radial (non-parallel) passages. These holes are also used to surround the turbine blades with a film of air which is at a lower temperature than the air entering the same section of the blade. A thermally insulating layer then effectively surrounds the blade which can operate in an ambient temperature higher than the melting point of the material from which it is manufactured.

The radial passage has normally been formed in some turbine blades, when they are cast, and the cooling holes for the air film produced by EDM or ECM. However some blades, especially those made from single crystal materials, have their radial passage machined out, usually because of difficulties in casting or forging them. Moreover, conventional or gun-drilling methods cannot effectively deal with these materials. Such blades may need holes with depth-to-diameter ratios as high as 160 to 1. Even EDM and laser machining have limitations in this respect and the ECM-based process, STEM drilling, was devised for this purpose.

As shown in Fig. 4.19, the configuration of electrodes is similar to that of ECM. The cathode drill-electrode is made of titanium tube, its outer wall having an insulating coating. A nitric acid electrolyte solution (about 15% v/v) at a temperature of about 20°C is pumped down the tube at a flow rate of approximately 1 $\text{l}\,\text{min}^{-1}$. With an applied voltage of 10 V, and a typical cathode-tool feed-rate of 2.2 $\text{mm}\,\text{min}^{-1}$, a hole of typical diameter varying from 0.84 mm at inlet to 0.85 mm at outlet, can be produced with a 0.58 mm diameter drill-tool through a depth of approximately 133 mm of blade. In STEM drilling periodic reversal of polarity, typically 9 s to 0.3 s, is employed to prevent an accumulation of undissolved machining products on the surface of the cathode-drill.

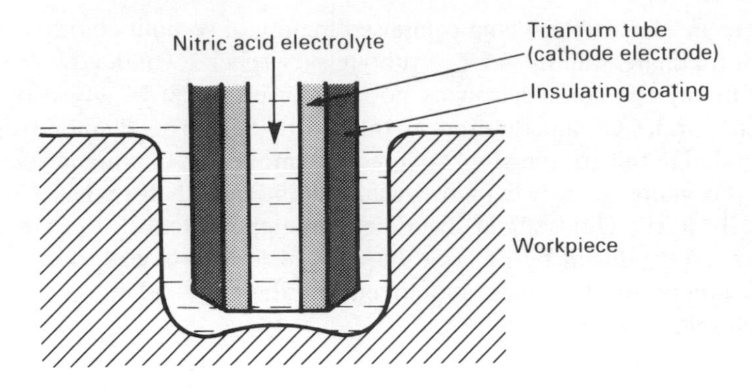

Fig. 4.19 STEM-drilling.

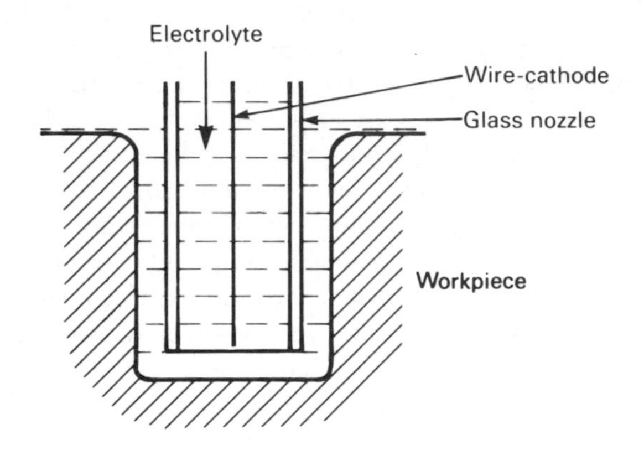

Fig. 4.20 Electrostream (capillary drilling).

4.11.3 Electrostream (capillary drilling)

Electrostream, also known as capillary drilling, is another special electrochemical machining technique, for production of fine holes. A schematic diagram of the equipment used is shown in Fig. 4.20. The cathode-tool is made from a drawn glass nozzle, 0.025 to 0.05 mm in diameter, smaller than the desired size of hole to be drilled. Within the glass nozzle a wire-electrode is fitted which provides the means for electrical contact with the acid electrolyte. The kind of electrolyte has to be carefully selected to ensure that it is chemically compatible with the condition of the workpiece. Generally, hydrochloric and sulphuric acid solutions are used, the former for anode materials such as aluminium, and related alloys, and the latter for Hastelloy,

Inco, Rene alloys, as well as 1010 carbon and 304 stainless steels (US notations). Voltages of about 120 V are employed, that is about ten times greater than those of ECM.

The drilling operation is conducted in a plastic chamber, suitably vented, with controls for the automatic rapid feed of the tool which has to match exactly the rate of recession of the workpiece area being machined. The tool is also fitted with a rapid retraction mechanism to enable indexing to the next location, for multiple holes to be drilled. The acid electrolyte is kept separated from the electrode feed mechanisms in order to protect the system from corrosion. Its temperature, pressure, concentration and flow rate are carefully monitored for satisfactory machining.

With electrostream, holes from 0.13 to 0.9 mm with depth-to-diameter ratios of up to 40 to 1 can be drilled. Holes can be drilled at angles as shallow as 10°.

A principal application is the drilling of cooling holes in nickel-based alloys. The technique also enables the drilling of cross-holes deep within a workpiece.

4.11.4 Electrochemical arc machining

A new process which relies on electrical discharges in electrolytes, thereby permitting metal erosion as well as ECM in that medium, has recently been developed. Since the new process relies on the onset of arcs rather than sparks, it has been named electrochemical arc machining (ECAM). (A spark

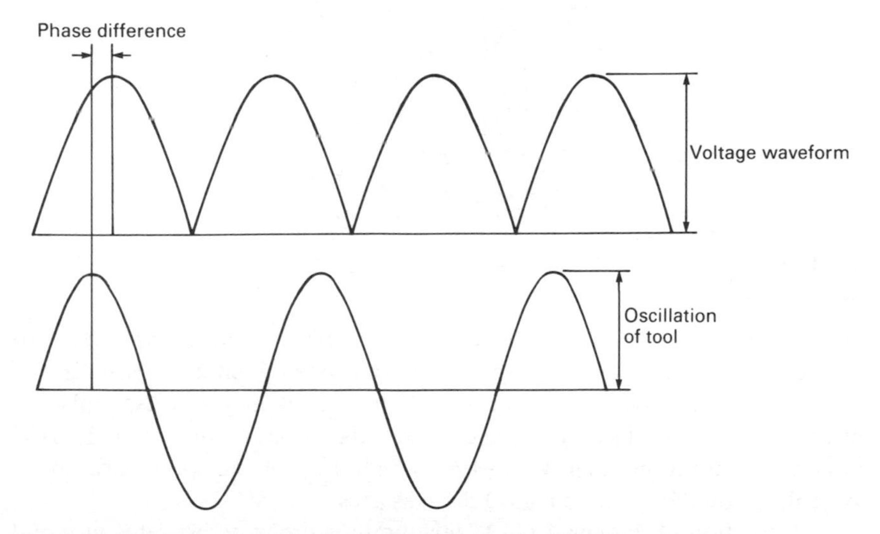

Fig. 4.21 Voltage and tool-oscillation waveforms for ECAM.

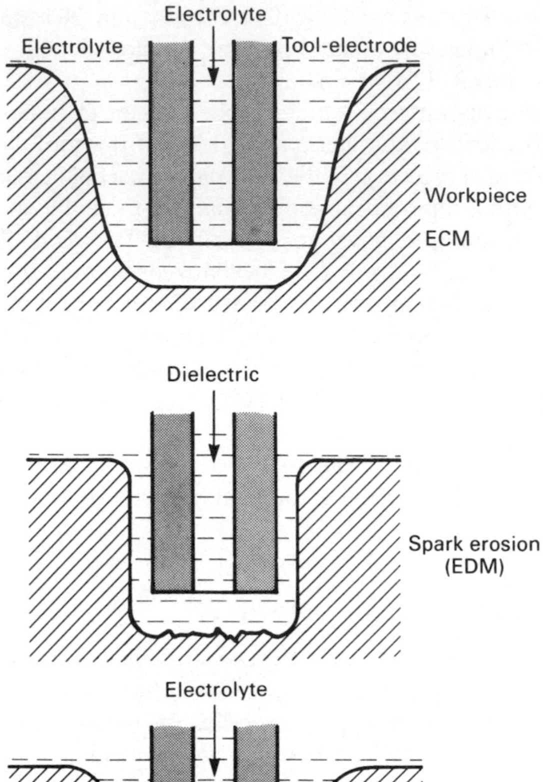

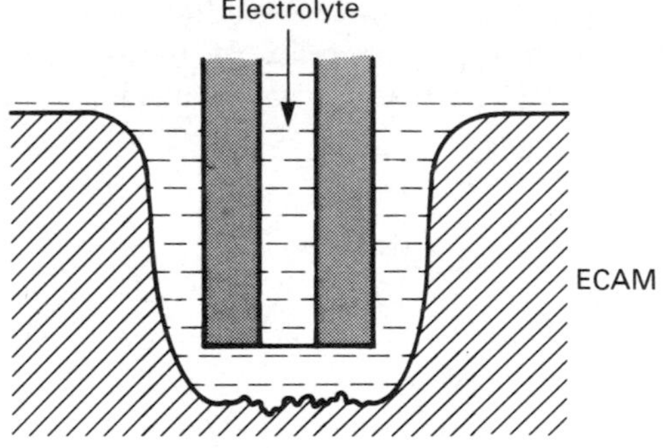

Fig. 4.22 Drilling by ECAM.

has been defined as a sudden transient and noisy discharge between two electrodes, whereas an arc is a stable thermionic phenomenon.) Discharges of duration approximately 1 μs to 1 ms could be described as sparks, whilst for durations of about 0.1 s they can be considered to be arcs. Since in the ECAM process the duration, energy and time of ignition of the sparks are under control, it would be valid to regard them as arcs.

An attraction of the new ECAM technique is the very fast rates of metal removal attainable by the combined effects of sparking and ECM. For example, in comparison with the rates for hole-drilling for EDM and ECM (respectively 0.1 and 5.0 mm min^{-1}), rates of 15 to 40 mm min^{-1} may be achieved by ECAM. The ECAM technique can be applied in all the ways discussed for ECM. Thus surfaces can be smoothed, and drilled; turning is possible, as is wire machining. (Khayry and McGeough, 1987; Elltofy and McGeough, 1988.)

One form of the new process relies on a pulsed direct current (i.e. full wave rectified a.c.) power supply which is locked in phase with a vibrating tool head. The oscillation of the tool at 100 Hz in phase with the pulsed d.c. power supply – a slight phase difference is permissible – gives rise to a set of conditions whereby ECM followed by electrical discharges causing spark erosion, and reverting back to ECM, occurs over each wave cycle. As the tool vibrates over one cycle, the inter-electrode gap narrows; during the same period the current rises until, for conditions of comparatively smaller gap and higher current, sparking takes place by breakdown of the electrolyte and/or generation of electrolytic gas or steam bubbles in the gap, the production of which aids the discharge process (Fig. 4.21).

For drilling, the discharge action occurs at the leading edge of the tool, whilst ECM still takes place on the side walls between the tool and the workpiece. The combined spark erosion and ECM action yields fast rates of metal removal. Since ECM is still possible, any metallurgical damage to the components caused by the sparking action can be removed by a short period of ECM (e.g. 15 s) after the main ECAM action. Currents of 250 A at 30 V are typically used in the process, which is illustrated in Fig. 4.22.

BIBLIOGRAPHY

Bellows, G. (1976) *Non-traditional Machining Guide 26 Newcomers for Production*, Metcut Research Associates Inc., Cincinnati, Ohio, pp. 28, 29.

Bellows, G. and Kohls, J. D. (1982) Drilling Without Drills, *American Machinist*, 178–83.

Clifton, D., Midgley, J. W. and McGeough, J. A. (1987) *Proc. Inst. Mech. Eng.* (in press).

Crichton, I. M., McGeough, J. A., Munro, W. and White, C. (1981) Comparative Studies of ECM, EDM, and ECAM, *Precision Engineering*, **3** (3), 155–60.

De Barr, A. E. and Oliver, D. A. (Eds) (1968) *Electrochemical Machining*, MacDonald Press.

De Silva, A. and McGeough, J. A. (1986) Surface Effects on Alloys Drilled by Electrochemical Arc Machining, *Proc. Instn Mech. Engrs*, **200** (B4), 237–46.

Drake, T. and McGeough, J. A. (1981) Aspects of Drilling by Electrochemical Arc Machining, *Proc. Machine Tool Design and Research Conf.*, Macmillan, pp. 362–9.

Elltofy, H. and McGeough, J. A. (1988) Evaluation of an Apparatus for Electrochemical Arc Wire Machining, *Trans. Am. Soc. Mech. Eng., J. Eng. Industry* (in press).

Ghabrail, S. R. *et al.* (1984) Electrochemical Wirecutting, *Proc. 24th Int. Machine Tool Design and Research Conf.*, Department of Mechanical Engineering, University of Manchester in association with Macmillan Press, pp. 323–8.

Graham, D. (1982) Deburring-2: Electrochemical Machining, *The Production Engineer*, **61** (6), 27–30.

Jain, V. K. and Nanda, V. N. (1986) Analysis of Taper Produced in Side Zone During ECD, *Precision Engineering*, **8** (1), 27–33.

Kaczmarek, J. (1976) *Principles of Machining by Cutting, Abrasion and Erosion*, Peter Peregrinus, Stevenage, pp. 487–513.

Khayry, A. B. M. and McGeough, J. A. (1987) Stochastic and Experimental Analyses of Electrochemical Arc Machining, *Proc. Roy. Soc. A*, **412**, 403–29.

Kubota, M. (1975) On the Technological Potentialities of ECDM, *Mechanique*, **303**, 15–18.

Kubota, M., Tamura, Y., Omori, J. and Hirano, Y. (1978) Basic Study of ECDM – I, *J. Assoc. of Electro-Machinery*, **12** (23), 24–33.

Kubota, M., Tamura, Y., Takahashi, H. and Sugaya, T. (1980) Basic Study of ECDM – II, *J. Assoc. of Electro-Machinery*, **13** (26), 42–57.

Landolt, D. (1987) *Experimental Study and Theoretical Modelling of Electrochemical Metal Dissolution Processes Involving a Shape Change of the Anode*, 172nd Meeting of the Electrochemical Society, Honolulu, Abstract No. 544.

Mao, K. W. (1971) ECM Study in a Closed Cell System, *J. Electrochem. Soc.*, **118**, 1870–9.

Mao, K. W., Michell, A. L. and Hoare, J. P. (1972) Anodic Film Studies on Steel in Nitrate-Based Electrolytes for ECM, *J. Electrochem. Soc.*, **119** (4), 419–27.

McGeough, J. A. (1974) *Principles of Electrochemical Machining*, Chapman and Hall, London.

Newton, M. A. (1985) *Stem Drilling Update*, Tech. paper, Society of Manufacturing Engineering, Paper MR85-381.

Wilson, J. F. (1971) *Theory and Practice of Electrochemical Machining*, John Wiley & Son, London.

5 Laser machining

5.1 INTRODUCTION

Laser machining is based on principles that were only recently discovered. The word 'laser' is an acronym of *l*ight *a*mplification by *s*timulated *e*mission of *r*adiation. The process depends on the interaction of an intense, highly directional coherent monochromatic beam of light with a workpiece, from which material is removed by vaporization. The explanations of atomic phenomena given in the Appendix are a useful basis for understanding laser effects.

The laser phenomenon was first predicted in 1958 by Schawlow and Townes. They visualized a Fabry–Perot interferometer acting as a resonance cavity. As indicated in Fig. 5.1 this instrument consists basically of two plane highly parallel half silvered mirrors between which a monochromatic beam of light undergoes multiple reflections. The cavity between the mirrors would be filled with an amplifying medium, gas molecules excited to high energy levels. The mechanism by which amplification occurs is described in later sections 5.3 and 5.4.

Light is propagated in a direction parallel to the axis of the interferometer, being reflected by the mirrors back through the excited gas; in this way the light can be amplified, providing of course that it maintains the same phase at successive reflections. As shown in Fig. 5.1 the growth of the light wave through the medium is represented by the upper growing wave W_1 moving from mirror M_1 to the right in Fig. 5.1. On reflection at mirror M_2 it yields another growing wave W_2 which travels in the opposite direction. The mirror M_2 is also partially transmitting. Thus the light beam emerges through it, diverging from the axis at an angle $\delta\theta$, where

$$\delta\theta = \lambda / D \tag{5.1}$$

where D is the diameter of the mirror. For example, for $D = 10\,\mathrm{mm}$ and

$\lambda = 1\,\mu m$, $\delta\theta = 10^{-4}$ rad, or 22 s of arc. That is, the emergent light is highly collimated. By the use of appropriate lenses, the divergence of the beam can be controlled; its diameter can be focused over a wider range of areas. For example a ruby laser (section 5.5.4) is excited by a flash lamp, emitting 1 kJ of electrical energy in 1 ms. A laser beam of 3 J energy at 6934 Å having a cross-section of 5 mm, and a divergence of 10^{-3} rad is produced. The beam, on focusing, can provide power densities of 1 MWcm^{-2}.

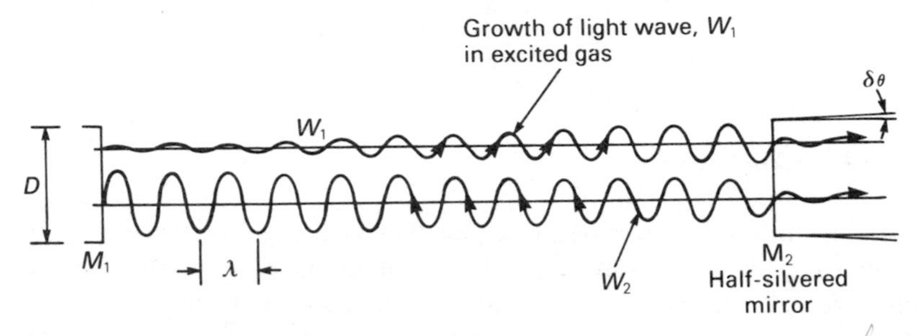

Fig. 5.1 Laser Fabry–Perot interferometer cavity.

The maximum directionality, or space coherence, of the beam set by the diffraction of the mirrors envisaged by Schawlow and Townes provoked immense interest from the outset. The means by which a highly collimated light beam could be achieved led to investigation of the physical phenomena associated with the emission of radiation. Spontaneous emission, a well-known phenomenon, is first considered.

5.2 SPONTANEOUS EMISSION OF RADIATION

When an atom in an excited state of energy E_i falls to a lower level E_j, it emits a quantum of radiation of frequency, ν_{ij}, where

$$E_i - E_j = h\nu_{ij} \tag{5.2}$$

where h is Planck's constant.

The same atom can be *stimulated* to emit this radiation if it receives radiation of the same frequency. The phenomenon of stimulation is less familiar than that of spontaneous emission. In view of its great relevance to laser work, it merits special attention.

5.3 STIMULATED EMISSION

The rate at which stimulated jumps in radiation occur is proportional to the energy density $u_{\nu_{ij}}$ of the radiation, and to the difference in the population (that is number per unit volume) of atoms between the upper and lower states. Both the stimulated and stimulating radiation have the same directional and polarization characteristics.

This process is the basis of the laser phenomenon. Further insight can be derived from work by Einstein on the interaction of matter and radiation (see Appendix). He proposed that a proper description of this interaction required the inclusion of conditions whereby an excited atom is induced by radiation in order to emit a photon which thereby decays to a lower energy state.

Suppose that a quantized atomic system has levels numbered 1, 2, 3 etc. with which are associated energies E_1, E_2, E_3. The number of atoms per unit volume (as noted above, the population) in these levels are N_1, N_2, N_3. If the atomic system is in equilibrium with thermal radiation at a temperature T, then the relative populations at any two levels, e.g. 1 and 2, are given by Boltzmann's equation:

$$\frac{N_2}{N_1} = \frac{e^{-E_2/kT}}{e^{-E_1/kT}} \tag{5.3}$$

where k is Boltzmann's constant.

If $E_2 > E_1$, then $N_2 < N_1$.

An atom can decay from levels 2 to 1, by emission of a photon. If A_{21} is the probability that a transition will occur per unit time for spontaneous emission from levels 2 to 1, then the number of spontaneous decays per second is $N_2 A_{21}$.

As well as these spontaneous jumps, induced or stimulated transitions will occur. The total rate of these induced transitions between levels 2 and 1 is proportional to the energy density u_ν of the radiation of frequency ν, where

$$\nu = \frac{(E_2 - E_1)}{h} \tag{5.4}$$

h being Planck's constant.

Let B_{21} and B_{12} be the proportionality constants for stimulated emission. The number per second of stimulated transitions, or emissions, in the downward direction is

$$N_2 B_{21} u_\nu \tag{5.5}$$

Similarly the number per second of stimulated transitions (absorptions) in the upward direction is

$$N_1 B_{12} u_\nu \tag{5.6}$$

The proportionality constants A and B are called the Einstein coefficients.

In the equilibrium, the net rates of transitions in the downward and upward directions are equal. Then

$$N_2 A_{21} + N_2 B_{21} u_\nu = N_1 B_{12} u_\nu \tag{5.7}$$

This equation can be arranged to give

$$u_\nu = \frac{N_2 A_{21}}{N_1 B_{12} - N_2 B_{21}} \tag{5.8}$$

From equation (5.3)

$$u_\nu = \frac{A_{21}}{B_{21}} \frac{1}{\dfrac{B_{12}}{B_{21}} \left\{ \exp\left(\dfrac{h\nu}{kT}\right) - 1 \right\}} \tag{5.9}$$

For this equation to agree with Planck's radiation law, given by

$$E = \frac{8\pi h\nu}{(e^{hc/kT\lambda} - 1)}$$

the following equations must hold:

$$B_{12} = B_{21}$$

$$\frac{A_{21}}{B_{21}} = \frac{8\pi h\nu^3}{c^3}$$

When atoms are in equilibrium with thermal radiation, the ratio of the rates of stimulated to spontaneous emission is given by

$$\frac{\text{stimulated rate}}{\text{spontaneous rate}} = \frac{B_{21} u_\nu}{A_{21}} = \frac{1}{(e^{h\nu/kT} - 1)} \tag{5.10}$$

On substitution of typical values for h and k, and for ordinary optical sources, where the temperature T is about 1000 K, the rate of induced emission is

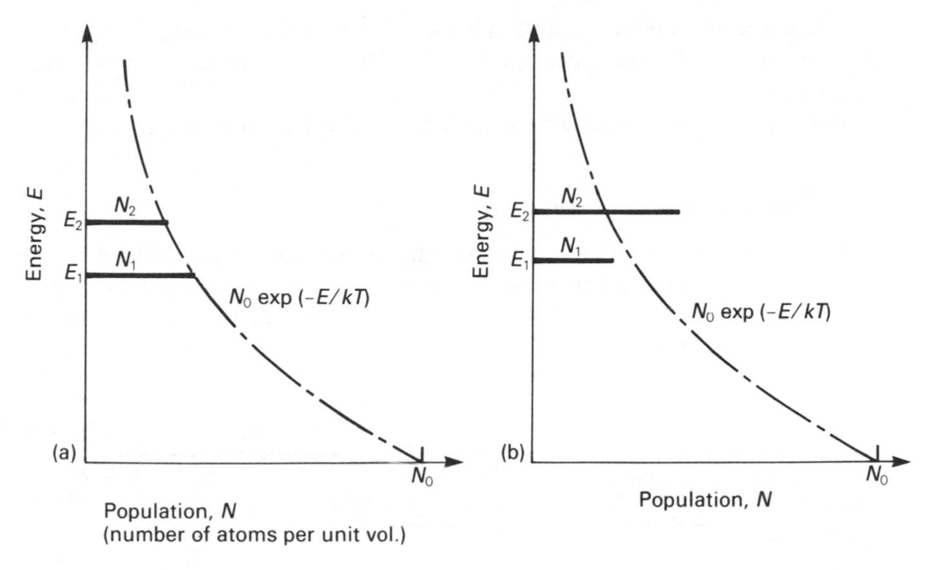

Fig. 5.2 Population densities. (After Fowles, 1975.)
(a) Boltzmann.
(b) Inverted distributions.

found to be very small in the visible part of the spectrum. That is, in these sources most of the radiation is emitted through spontaneous transactions, and since these occur in a random fashion ordinary sources of visible radiation are incoherent.

In comparison, in a laser the radiation density builds up such that induced transitions become completely dominant, and the emitted radiation is very coherent. Moreover, the spectral radiation of the laser at its operating frequency is much greater than that of ordinary light. A new condition, called *population inversion*, is needed to obtain this effect with lasers.

Consider an optical medium which contains atoms at various energy levels E_1, E_2, E_3 etc., with $E_1 < E_2$. The rates of stimulated emission and absorption involving these two levels are proportional, respectively, to $N_2 B_{21}$ and $N_1 B_{12}$. Since $B_{21} = B_{12}$, the rate of stimulated downward transitions is greater than those in the upward direction, if $N_2 > N_1$. That is, provided that the population of the upper state exceeds that of the lower one. This condition is illustrated in Fig. 5.2 (a) and (b). It is termed population inversion, and clearly contravenes the Boltzmann thermal equilibrium distribution (cf. equation 5.3).

If the population inversion exists, the intensity of a light beam can be shown to increase as it traverses the medium. That is, the beam will be amplified, since the gain due to the induced emission exceeds the loss due to absorption.

The induced radiation is emitted in the same direction as the primary beam. The two have a definite phase relationship. That is, the induced and primary radiations are *coherent*.

Population inversion may be achieved by several methods including

(a) Optical pumping (e.g. ruby laser)

As shown in Fig. 5.3(a) an external light source is used to produce a high population of some particular energy level in the laser medium, by selective optical absorption. Optical excitation of this type is used in solid-state lasers, such as the ruby type.

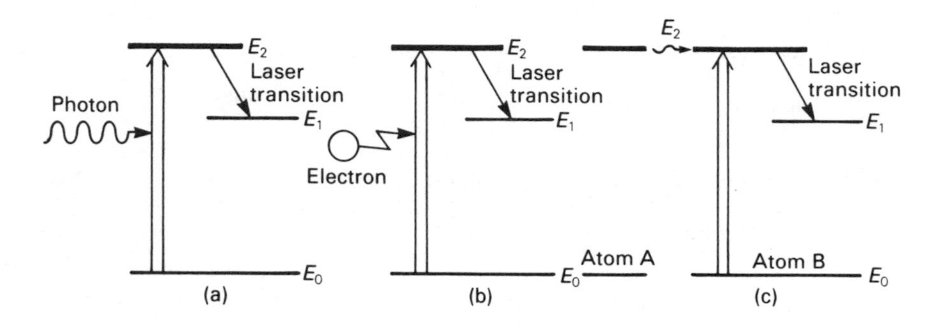

Fig. 5.3 Production of population inversion. (After Fowles, 1975.)
(a) Optical pumping.
(b) Direct electron excitation.
(c) Inelastic atom–atom collisions.

(b) Direct electron excitation (e.g. argon laser)

In gaseous ion (e.g. argon) lasers, direct electron excitation in a gaseous discharge is used to produce the population inversion. The laser medium itself carries the discharge current. For the appropriate conditions of pressure and current, the electrons in the discharge may directly excite the active atoms to produce a higher population in certain levels, compared to lower ones. This is illustrated in Fig. 5.3(b).

(c) Inelastic atom–atom collisions (e.g. helium–neon laser)

This type is shown in Fig. 5.3(c). A combination of gases is used, such that two different types, e.g. A and B, both have some excited states A* and B* that coincide (or nearly so). The transfer of excitation may occur between the two atoms such as

$$A^* + B \rightarrow A + B^*$$

If the excited state of one of the atoms, e.g. A*, is metastable, then the presence of gas B will serve as an outlet for the excitation. As a result, the excited level of atom B may become more highly populated than some lower level to which the atom B can decay by radiation.

This condition arises with the helium–neon laser. An excited helium atom excites a neon atom, and the laser transition then occurs in the latter.

5.4 LASER OSCILLATION

The amplifying medium for a laser is located between two mirrors fixed at either end of a tube. If sufficient population inversion occurs in the medium, then the electromagnetic radiation builds up, and becomes established as a standing wave between the mirrors. One of the mirrors is partially transmitting. When the radiation becomes sufficiently strong, it emerges through the latter mirror as a well-defined laser beam.

5.5 TYPES OF LASER

Several laser types are of interest for machining: gas and solid-state lasers are particularly noteworthy.

5.5.1 Gas laser

Figure 5.4 shows a typical arrangement for the gas laser. The ends of the laser tube are fitted with highly transparent 'Brewster' windows by means of which the output of the laser is linearly polarized. Laser oscillation at the preferred

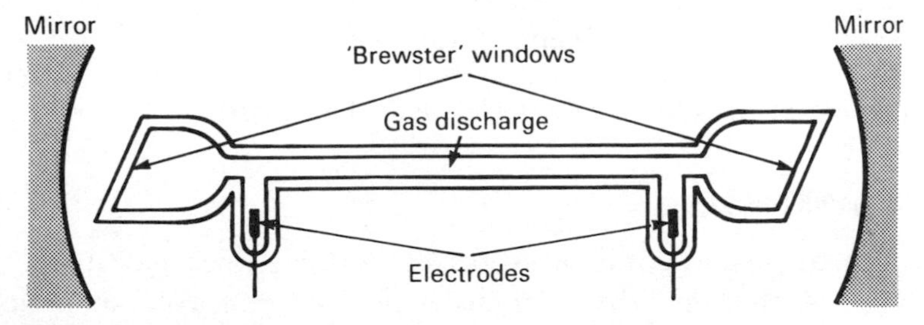

Fig. 5.4 Basic elements of gas laser.

polarization builds up, and becomes dominant over the other (orthogonal) polarization.

External electrical excitation can be obtained from several sources, such as direct and alternating current discharges. The latter is very simple: the power source can be an ordinary high voltage transformer to which are connected cold metal electrodes in the tube.

An electrodeless high-frequency discharge was used in the helium–neon laser, the first gas laser, developed by Javan, Bennett and Herriot at the Bell Telephone Laboratories. Many high-powered lasers use high voltage pulses, especially when steady population inversion cannot be maintained.

5.5.2 Helium–neon laser

The energy levels for this common gas laser are shown in Fig. 5.5. Helium atoms are excited by electron impact in the discharge. The populations of the helium metastable states denoted by states ^3S and ^1S build up, because there are no optically allowed transitions to lower levels. As shown in Fig. 5.5 the neon levels 2s and 3s lie close to the metastable helium levels.

Energy transfer is therefore highly probable when a metastable helium atom collides with an unexcited neon atom. These energy transfers may be expressed as:

$$\mathrm{He}\ (^3\mathrm{S}) + \mathrm{Ne} \rightarrow \mathrm{He} + \mathrm{Ne}\ (2\mathrm{s})$$

$$\mathrm{He}\ (^1\mathrm{S}) + \mathrm{Ne} \rightarrow \mathrm{He} + \mathrm{Ne}\ (3\mathrm{s})$$

When the discharge conditions are suitable – the total pressure should be about 1 Torr, and the best ratio of helium to neon is about 7 to 1 – population inversion of the Ne(2s) and Ne(3s) levels can occur.

The main laser action in the helium–neon system corresponds to the following transitions in the neon atom.

$$3s_2 \rightarrow 2p_4 \qquad (632.8\ \mu m)$$

$$2s_2 \rightarrow 2p_4 \qquad (1.1523\ \mu m)$$

$$3s_2 \rightarrow 3p_4 \qquad (3.39\ \mu m)$$

The numbers in brackets give the wavelength of the laser.

5.5.3 Carbon dioxide laser

Very high continuous power levels became possible when the CO_2:N_2:He laser was devised in the 1960s, with the output rising from a few milliwatts to hundreds of watts.

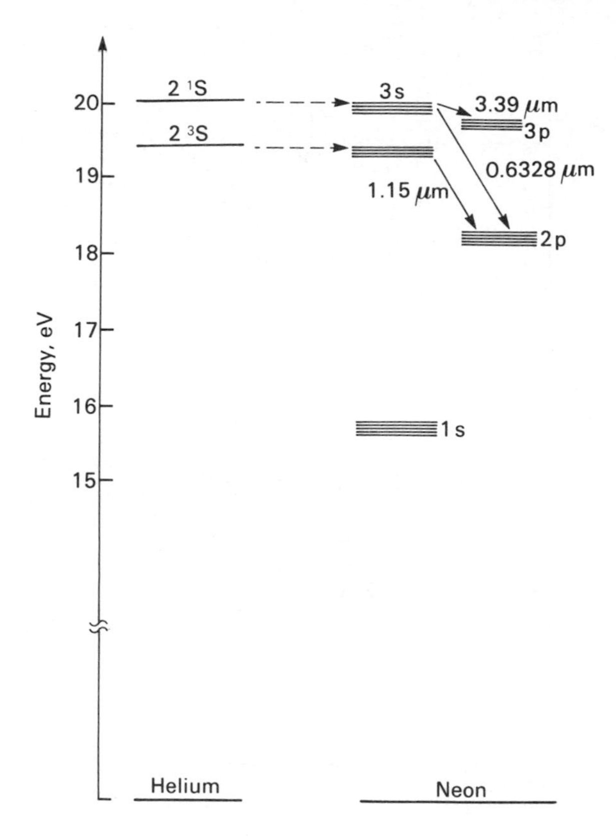

Fig. 5.5 Energy levels for helium–neon laser.

From the energy level diagram in Fig. 5.6, the laser transition occurs, between two (00°1 and 10°0) vibrational energy levels of the carbon dioxide molecules, the difference between which corresponds to an output wavelength of 10.6 μm. Detailed studies have revealed that the nitrogen molecules in the discharge are excited to a vibrational level ($v = 1$) which is very close to the 00°1 level in the CO_2. This excitation in N_2 is effectively transferred to the upper laser level of the CO_2, creating an excess of molecules there. This preferential population is a necessary condition for the occurrence of lasing action. The helium in the discharge helps to maintain the population inversion, as well as improving heat conduction to the walls.

After the lasing action at a wavelength of 10.6 μm, the CO_2 molecule decays to the 01^10 level, and then radiates to the ground state. Figure 5.7 illustrates the main features found in commercially available CO_2:N_2:He lasers. Figure 5.8 shows an industrial CO_2 laser machine tool. Table 5.1 gives

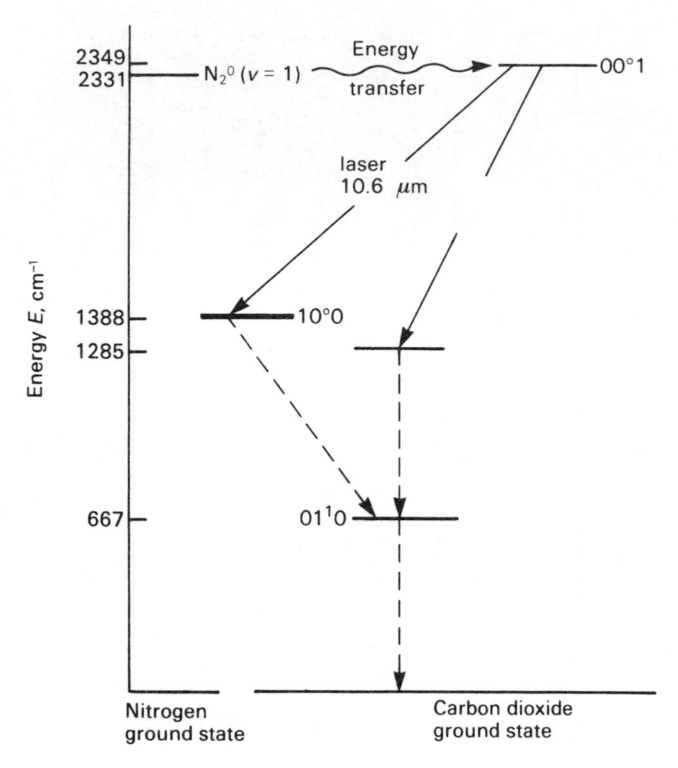

Fig. 5.6 Energy levels for nitrogen and carbon dioxide molecules. (After Weaver, 1971.)

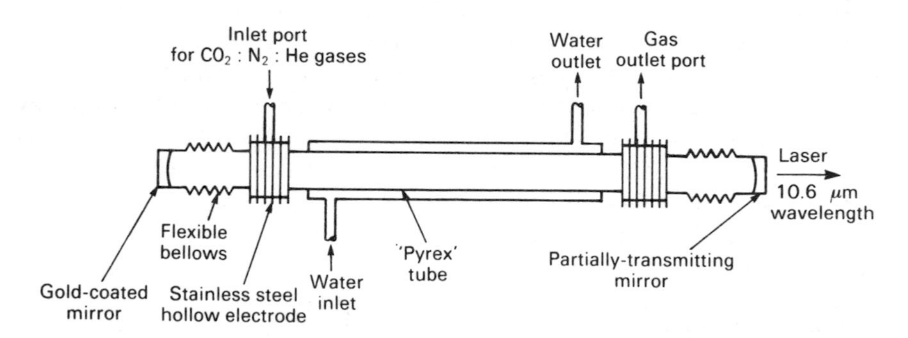

Fig. 5.7 Main features of CO_2: N_2: He laser.

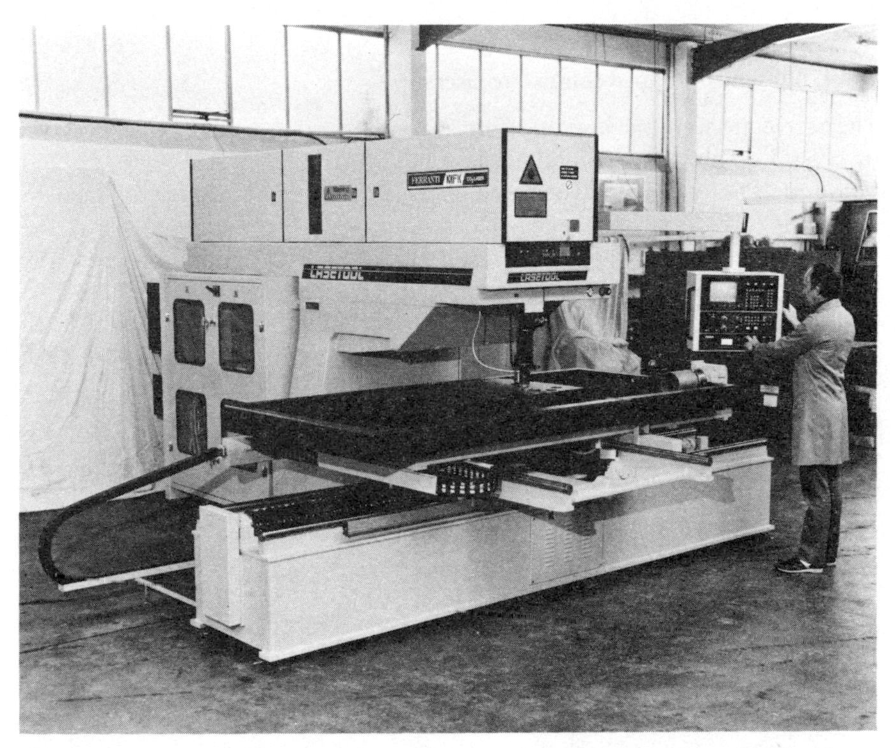

Fig. 5.8 Laser cutting machine. (Courtesy of Ferranti Industrial Electronics Ltd.)

Table 5.1 Types of gas lasers (Data from Fowles, 1975)

Gas or gas mixture	*Active species*	*Principal laser wavelength (μm)*	*Other information*
HeNe	Ne	0.6328, 1.15, 3.39	continuous wave (cw)
Ar	Ar^+	0.4765, 0.4880, 0.5145	cw or pulsed (p)
CO_2–N_2–He	CO_2	10.6	pulsed and cw, high power, high efficiency

some relevant information on the properties of gas lasers of interest in machining.

5.5.4 Optically pumped solid-state lasers

The active atoms of the laser medium are embedded in a solid, typically a rod of crystal or glass, with parallel, flat ends which are optically ground and polished. The rod may have coated ends to form the optical cavity needed; alternatively external mirrors can be used.

Xenon-flash and high-pressure mercury-discharge lamps are the usual external light sources used to produce the optical pumping of the active atoms. A common configuration of solid-state lasers is shown in Fig. 5.9. In Fig. 5.9(a) the laser rod, with coated ends, is housed within a helical flash lamp; in Fig. 5.9(b) the flash lamp has a straight form.

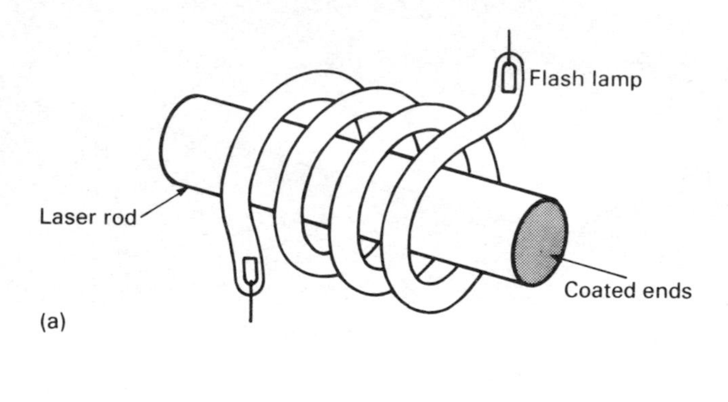

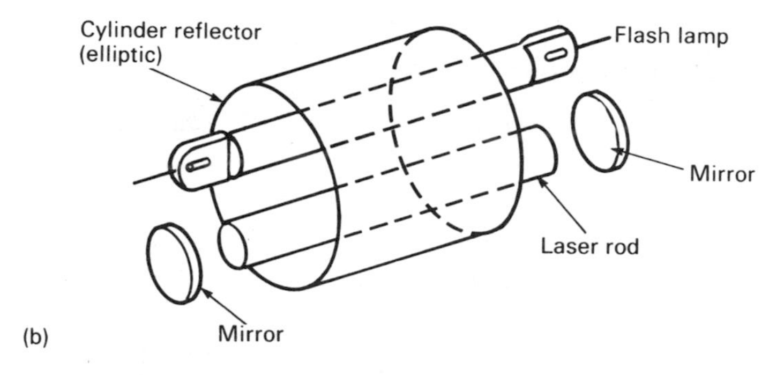

Fig. 5.9 Optically pumped solid-state lasers.
(a) Helical flash lamp.
(b) Straight flash lamp.

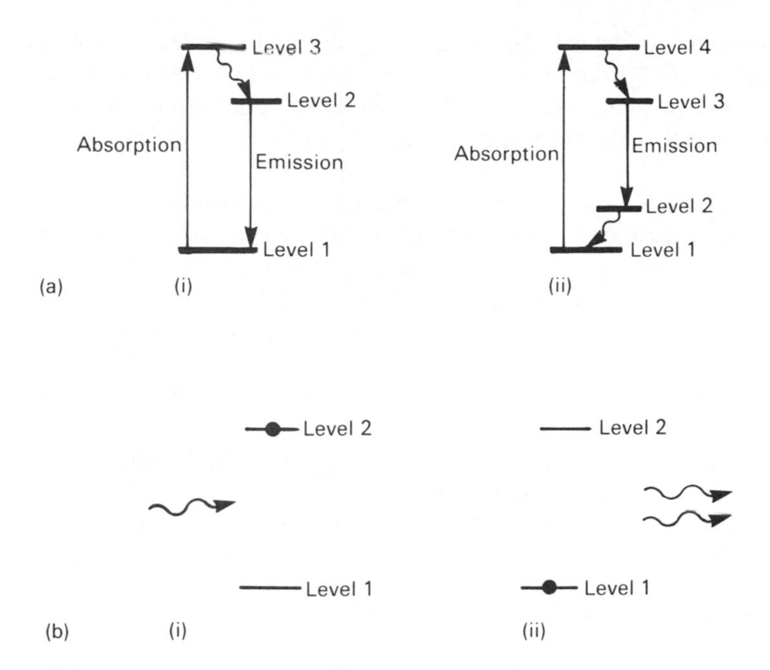

Fig. 5.10
(a) Absorption and emission.
 (i) Three-level laser.
 (ii) Four-level laser.
(b) Conditions for stimulated emission.
 (i) Before stimulated emission.
 (ii) After stimulated emission.

The ruby laser is representative of this type. Its main component is a rod made of synthetic crystalline sapphire (Al_2O_2) doped with approximately 0.05% by weight of chromium oxide (Cr_2O_3). The aluminium in the crystal lattice is replaced by chromium (Cr^{+++}) ions which endow the material with a pink hue. The ruby crystals can be grown in lengths of about 300 mm with a diameter of 25 mm. The material so produced is hard and durable, and of good optical quality; it has a high thermal conductivity so that the laser rod can be readily cooled (which is needed to maintain the operation of the laser).

As illustrated in Fig. 5.10 (a) the laser action is obtained by absorption of the pumping light by the chromium (Cr^{+++} ions) raising them from the ground state to a band of energy levels above the upper laser level (level 3 in Fig. 5.10(a)). Rapid non-radiative decay takes place to the upper laser level 2. The lasing takes place to the ground state Fig. 5.10(b) providing an excess of ions exists in the upper laser level. For this to occur more than 50% of the ions in the ground state have to be excited, in order that population inversion with respect to the ground state can occur. This three-level-type lasing action requires high pumping, normally provided by pulsing action. The typical output wavelength and power of the ruby laser are respectively 0.6943 μm and 400 J.

Some solid-state lasers operate on a four-level system. Here the ground state ions are excited by the pumping light to level 4 (Fig. 5.10). They then decay to the upper laser level 3. Laser action occurs to level 2 if there is a population inversion between these levels, and the ion is de-excited to the ground state. Only a small percentage of the ground state ions needs exciting to provide the population inversion between levels 3 and 2; cf. the three level system which needs more than 50%. Consequently four level systems have much lower pumping needs and laser thresholds.

A well-known member of four level group is the Nd^{3+} ion, which is a key constituent of the YAG laser (crystalline $Y_3A1_5O_{12}$ doped with Nd^{3+}). Table 5.2 shows some of the main physical properties and performances of solid-state lasers.

Table 5.2 Properties of solid-state lasers (after Weaver, 1971)

Material (active ion)	*Output wavelength (μm)*	*Mode of operation*	*Output (J)*
Ruby (Cr^{3+})	0.6943	Pulsed	500
YAG (Nd^{3+})	1.06	Continuous	1100 W
		Pulsed	10 J
		Q-switched	2×10^7 W
Glass (Nd^{3+})	1.06	Pulsed	5000 J

5.6 LASER BEAM CHARACTERISTICS

From the discussions above, a laser differs from ordinary light in terms of the following:

1. 'spatial profile'
2. beam divergence
3. focusing
4. temporal behaviour
5. brightness
6. power.

For laser machining, characteristics 1, 2, 3 and 6 are most significant.

5.6.1 Spatial profile

Lasers have a characteristic spatial pattern called Transverse Electromagnetic Modes (TEM). Briefly the transverse mode determines the propagation and focusing of the beam. The TEM are a consequence of resonance within the laser cavity, and are a measure of the configurations of the electromagnetic field determined by the boundary conditions in the cavity. The subscripts 'mm' are often used to refer to the number of nulls in the spatial pattern that occur in each of two orthogonal directions transverse to the direction of beam propagation.

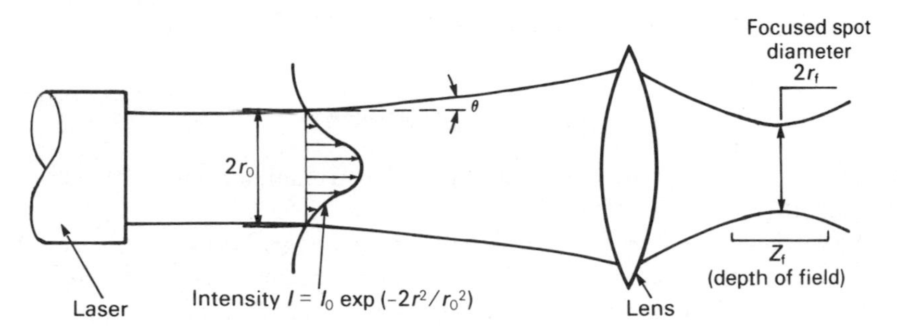

Fig. 5.11 Conditions for TEM_{00} mode.

The TEM_{00} mode is often used for machining. As illustrated in Fig. 5.11, the intensity of the laser beam in this mode follows a Gaussian distribution as a function of radius r, from the centre of the beam:

$$I(r) = I_0 \exp(-2r^2/r_0^2) \tag{5.11}$$

where I_0 is the intensity of the beam at the centre, and r_0 is the radius at which the intensity is reduced from its central value by the factor e^2. (Sometimes r_0 is called the 'Gaussian' radius.)

5.6.2 Beam divergence

The laser beam is highly directional and collimated. Thus the energy that it carries can be readily focused to a small area. From diffraction theory, the lower limit of beam divergence θ_t is given by

$$\theta_t = \frac{K\lambda}{d} \tag{5.12}$$

where $K = 2/\pi$ for a Gaussian beam, d is the aperture diameter through which the beam emerges, and λ is the wavelength of the beam.

For a laser operating in the fundamental mode, the beam divergence is typically 0.001 rad. A high-powered laser operating in a multi-mode pulsed fashion will have an angle of divergence of 0.015 to 0.020 rad.

5.6.3 Focusing and power

The diameter of the unfocused laser beam can be several mm wide. Focusing is needed to provide sufficient power density, so that the temperature of the materials to be treated is raised above the melting or boiling point. The diameter d_f of a Gaussian beam, focused by a simple lens, is given by

$$d_f = \frac{2f\lambda}{\pi d} = f\theta \tag{5.13}$$

where f is the focal length of the lens, d is the beam diameter, and λ is the laser wavelength.

The smallest minimum spot sizes for higher-mode, gas lasers, which cannot be focused to the diffraction limit can be about 0.1 mm. When the radiant energy of a laser is focused by a lens, the power density P at the focal plane of the lens can often be represented by the expression:

$$P = \frac{4E}{\pi f^2 \theta^2 t} \tag{5.14}$$

where E is energy output from the lens, f is the focal length, θ is full angle beam divergence, and t is the laser pulse length.

The minimum beam divergence θ_{min} of a light beam is a function of its wavelength and beam diameter. The minimum beam divergence of a spatially coherent beam can be deduced, from the Rayleigh criterion, as

$$\theta_{min} = \frac{1.22\lambda}{R} \tag{5.15}$$

where λ is the wavelength, and R is the radius of the beam or aperture. With higher energy lasers, the minimum beam divergence can be within a factor of three of that predicted by the Rayleigh criterion: a half-angle beam divergence of 2 to 10 mrad can be obtained.

Example

Suppose the beam divergence of the laser is 5 mrad, and that its radiation is focused with a 25 mm focal length lens. Then the power density within the focal spot is 2 000 times the power within the unfocused laser beam. Typically, the peak powers obtained from pulsed laser systems range from 10 000 W to 1 MW. The focused beam would then have a power density of approximately 20 MWcm^{-2} to 2 GWcm^{-2}. Even if only 0.1% of this energy is absorbed by a material, the rate of power absorption would be 20 kWcm^{-2} to 2 MWcm^{-2}.

In a typical pulsed laser, the focused spot size S can be estimated from

$$S = f\theta \tag{5.16}$$

Example

From the above values of beam divergence and lens focal length, a diameter for the focal spot of 0.25 mm is obtained. Much smaller values can be obtained, by use of lenses of shorter focal length and improved laser beam divergence.

5.7 EFFECTS OF LASER ON MATERIALS

When the laser beam meets the workpiece, as illustrated in Fig. 5.12, several effects arise, including reflection, absorption, conduction of the light. A useful account of these main effects by Weaver (1971) may help towards understanding laser machining.

5.7.1 Reflectivity

The amount by which the beam is reflected depends on the wavelength of the laser radiation, and on the condition and properties of the material, such as its surface finish, the amount to which it is oxidized, and its temperature. In particular, the high reflectivity of many materials at certain laser wavelengths renders them unsuitable for machining. Generally, the longer the wavelength of the laser beam, the higher becomes the reflectivity of metals. This observation is confirmed by the results in Table 5.3 which shows in particular the high reflectivity obtained with the 10.6 μm wavelength of the CO_2 laser. For wavelengths greater than 5 μm, most metals reflect about 90% of the incident radiation at low power densities.

The amount of reflectivity can be substantially reduced by modification of the surface condition of the workpiece. For example, the reflectivity of copper at a wavelength of 694.3 nm has been reported to be reduced from 95% to less than 20% by oxidizing the surface. (Note: although data on reflectivities are useful for evaluating laser effects on smooth uncontaminated surfaces, they should be used with caution for interpreting other cases where these conditions may not necessarily apply.)

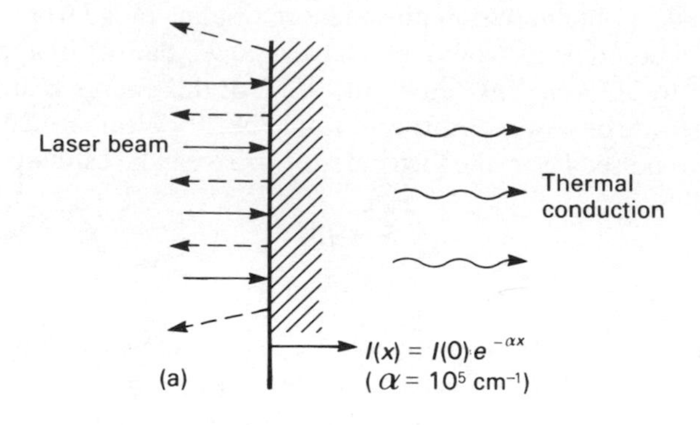

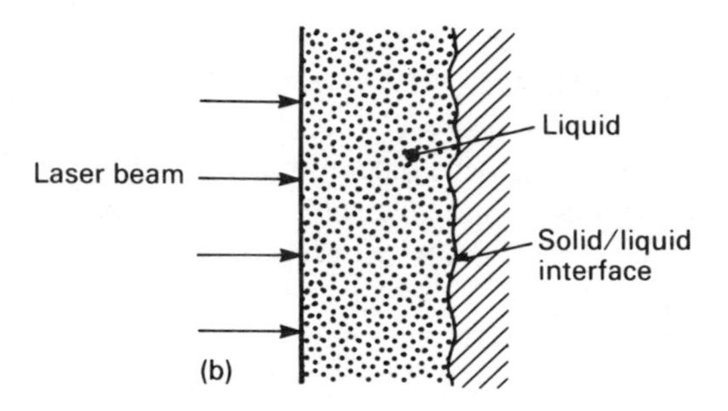

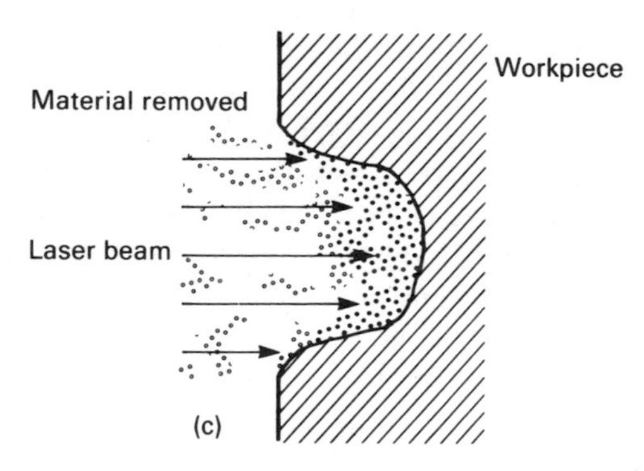

Fig. 5.12 Interaction of laser beam with workpiece.
(a) Absorption and heating.
(b) Melting.
(c) Vaporization.

Table 5.3 Reflectivity of metals at normal incidence of laser light (Data from Weaver, 1971)

Wavelength (*μm*)	*Au*	*Cu*	*Ag*	*Al*	*Cr*	*Fe*	*Ni*
0.6943 (Cr^{3+})	0.930	0.831	0.961	–	0.555	0.575	0.676
1.06 (Nd^{3+})	0.981	0.901	0.964	0.733	0.570	0.650	0.741
10.6 (CO_2)	0.975	0.984	0.989	0.970	0.930	–	0.941

Nonetheless reduction in reflectivity leads to increase in absorption of the laser energy by the surface, with subsequent effects on the material. Next, therefore, absorption is examined.

5.7.2 Absorption

Laser energy which is not reflected at the surface is absorbed into the material. The absorption of the light in metals takes place by an internal photo-electric effect which raises the electrons to higher energy states in the conduction band of the metal. (See, for example, Pascoe (1973) in Appendix.) Now the mean free time between collisions for electrons in a conductor is of the order of 10^{-14} to 10^{-13} s. Thus in 1 nanosecond (ns), the electrons will have made 10^{14} to 10^{15} collisions among themselves. Since this is a very short period compared to even the shortest laser pulse, the energy absorbed by the electrons from the laser beam is rapidly passed to the lattice.

The depth over which the absorption takes place may be approximated from the mean free path of the valence electrons, according to the relation

$$I(x) = I(0)e^{-\alpha x}$$

where $I(x)$ is the light intensity (in W) at depth x of penetration into the material, $I(0)$ being the incident intensity, and α is an absorption coefficient (m^{-1}).

Most of the energy is found to be absorbed in a 'skin depth' S given by

$$S = \alpha^{-1} \tag{5.17}$$

Typically the energy is absorbed in a depth of about 0.1 μm (for visible and infra-red wavelengths). For most organic compounds, absorption is found to take place in less than 1 μm (for CO_2 for infra-red radiation).

In summary, the laser energy may therefore be regarded as a surface effect, with the energy penetrating further into the material by thermal conduction.

5.7.3 Heat conduction

The conduction of the heat from the laser into the workpiece material is an extremely complex effect. As a result no adequate theory of heat conduction has yet been applied to laser machining. Nonetheless, useful information relevant to laser machining can be derived from a simple approach.

Firstly, since the workpiece is assumed to be composed of an isotropic material, the heat flow through it can be described by the diffusion equation

$$\frac{\partial T}{\partial t} = \kappa \nabla^2 T \tag{5.18}$$

Here T is absolute temperature (K) and t is time (s). κ, the diffusivity, is given by

$$\kappa = k/\rho c$$

k is the coefficient of thermal conductivity ($\mathrm{Wm^{-1}K^{-1}}$), ρ is the density ($\mathrm{kgm^{-3}}$), c the specific heat ($\mathrm{Jkg^{-1}K^{-1}}$) of the solid material.

Useful information about the way the heat from the laser is spread through the workpiece can be deduced from the solution to the diffusion equation (5.18) on the assumptions that the distribution of the temperature occurs in one dimension and that the workpiece can be regarded as semi-infinite in length.

For conditions of zero initial temperature throughout the material, and with the surface $x = 0$ maintained at T_0 for time t, where $t > 0$, the solution from Carslaw and Jaeger (1959) is

$$T(x,t) = T_0 \operatorname{erfc}(x/2\sqrt{\kappa t})$$

where the error function 'erfc' is given by

$$\begin{aligned} \operatorname{erfc} z &\equiv 1 - \operatorname{erf} z \\ &= 2/\sqrt{\pi} \int_0^\infty \mathrm{e}^{-\xi^2}\, d\xi \end{aligned}$$

$$\operatorname{erf} z = 2/\sqrt{\pi} \int_0^z \mathrm{e}^{-\xi^2}\, d\xi$$

Table 5.4 Theoretical times of penetration for isotherm ($T/T_0 = 0.5$) at various depths into various metals (Data from Weaver, 1971)

Material	*Thermal diffusivity ($cm^2 s^{-1}$)*	*Penetration time (s)*		
		x = 0.1	*x = 1.0*	*x = 10 cm*
Aluminium	0.91	0.0122	1.22	122
Chromium	0.20	0.0557	5.57	557
Copper	1.14	0.0098	0.98	97.8
Gold	1.18	0.0095	0.95	95
Silver	1.71	0.0065	0.65	65
Titanium	0.082	0.136	13.6	136

Example

For conditions of temperature where the ratio $T(x,t)/T_0 = 0.5$, that is an increase in the temperature to one-half of that at the wall of the workpiece, and from tabulated data on error functions

$$x/2\sqrt{\kappa\tau} = 0.523$$

The isotherm defined by the ratio $T(x,t)/T_0$ would require a time

$$t = 1.114/\kappa$$

to penetrate to a depth of 1 cm into the workpiece. Table 5.4 has been compiled to show the penetration times for a range of common materials. Even materials of high thermal conductivity have penetration times of less than 1 s for a depth of 1 cm. In comparison chromium and titanium need about 10 s.

5.7.4 Melting of surface

On sufficient heating by the laser the workpiece starts to melt. Again results from Carslaw and Jaeger are useful. The temperature rise due to a heat flux incident upon the surface ($x = 0$) of the workpiece, still assumed to be a semi-infinite medium, can be calculated as

$$T(x,t)|_{x=0} = [2F_0/k](\kappa t/\pi)^{1/2} \quad (5.19)$$

where F_0 is the constant heat flux ($Js^{-1}cm^{-2}$).

From the data on the properties of the material and power of the incident laser beam, Table 5.5 shows the times calculated from equation (5.19) needed to achieve surface melting.

Table 5.5 Theoretical time for onset of surface melting of metals subjected to constant heat flux (Data from Weaver, 1971)

Material	*Thermal conductivity* $Wcm^{-1}K^{-1}$	*Diffusivity* cm^2s^{-1}	*Melting temperature* K	*Melting time (s) at an incident flux* 10^4 Wcm^{-2}	10^6 Wcm^{-2}
Aluminium	2.38	0.91	933	4.24×10^{-2}	4.24×10^{-6}
Chromium	0.87	0.20	2176	1.4×10^{-1}	1.4×10^{-5}
Copper	4.0	1.14	1356	2.03×10^{-1}	2.03×10^{-5}
Gold	3.11	1.18	1336	1.15×10^{-1}	1.15×10^{-5}
Silver	4.18	1.71	1234	1.22×10^{-1}	1.22×10^{-5}
Titanium	0.20	0.082	1941	1.45×10^{-2}	1.45×10^{-6}

Example

For an incident flux of 10^4 Wcm^{-2}, the surface melting is obtained in 0.1 s for chromium, and 0.01 s for titanium.

Melting time is inversely proportional to the square of the incident power density. Thus at a power density of 10^6 Wcm^{-2}, melting times are reduced to between 10^{-5} and 10^{-6} s. Thus even if only 1% of the incident power is absorbed on the surface, surface melting can still occur in less than about 0.1 s with beam densities of 10^6 Wcm^{-2}.

When the heat fluxes associated with the laser are below about 10^4 Wcm^{-2}, equation (5.19) is generally held to be a reasonable representation of temperature rise in the workpiece. For heat fluxes greater than 10^6 Wcm^{-2}, and when the radius of the region being heated is less than about 0.1 cm, more complicated expressions have to be used (see Weaver, 1971).

5.7.5 Vaporization

Very rapidly after melting by the laser, vaporization of the workpiece surface commences. The rate of vaporization may be related to the incident flux F of the laser by the expression

$$F = \left(\frac{dx}{dt}\right)C \qquad (5.20)$$

where (dx/dt) is rate of recession of the workpiece surface, and C is the energy needed to vaporize unit volume of the workpiece. Typically C is about 10^3 Jcm^{-3}. (Note: the use of this simple equation implies that vaporization does not interfere with the laser beam as it meets the workpiece.)

Example
An incident laser flux of 3×10^8 Wcm^{-2} would enable the workpiece surface to be vaporized at a recession rate of approximately 10^{-4} cms^{-1}. Now at such high intensities of the laser beam the incident energy has already been noted to be absorbed in the so-called skin depth. In section 5.7 this depth was observed to be less than 1 μm thick. Under these conditions, absorption into the bulk of the solid can be neglected, and very rapid vaporization of the workpiece material can be expected.

The time t_0 taken for vaporization to start can be estimated from equation (5.20) as approximately given by:

$$t_0 = SC/F \tag{5.21}$$

where S is the skin depth over which the energy from the laser is absorbed.

Example
For S even as thick as 10^{-4} cm, and $F = 3 \times 10^7$ Wcm^{-2}, the time t_0 is found to be about 10^{-8} s. That is, vaporization occurs very rapidly at these high incident laser fluxes. An assumption relating to equations (5.20) and (5.21) should be noted: these equations imply that the vaporization itself does not interfere with the laser beam as it strikes the surface. However, in detailed analyses the rapidity with which vaporization occurs requires that the effects of energy absorption by the vapour should be taken into account. That is, the vapour does really interfere with the incident laser beam.

5.7.6 Energy needed for vaporization

In the light of the above discussion, Table 5.6 is a useful guide to the minimum laser energies needed to melt and to vaporize a wide range of common materials.

The table indicates the comparatively low energy needed to vaporize plastics, compared with metals. Radiation of the wavelength of the CO_2 laser (10.6 μm) is readily absorbed by most non-metals, which also usually have low thermal conductivity. Thus, these plastic materials can be readily melted by low power (several watts) CO_2 lasers. They can be cut at high speeds with slightly higher powers. For instance, a 400 W CO_2 laser can cut through 0.1 mm thick plastic at a rate of more than 4 ms^{-1}.

Since metals have a higher reflectivity and thermal conductivity than plastics, greater power densities are usually needed to cut them. A method for estimating the energy needed for vaporization is useful. An illustrative calculation of this kind, carried out by Gagliano *et al.* (1969) is presented below.

Table 5.6 Minimum laser energy needed to melt and vaporize various materials

	Energy required (kJcm^{-3})	
Melting of metals	2.5 (Aluminium)	to 12.5 (Tungsten)
Vaporization (decomposition) of quartz, wood, acrylic plastic	2	to 4
Vaporization of metals	30	to 80

Example

Suppose a laser has to produce 10 μm diameter hole in a gold film of thickness 0.5 μm. The volume and mass to be evaporated are calculated to be respectively 39.25 μm^3 and 7.58×10^{-13} kg, for a density of gold of 19 300 kgm^{-3}.

Next the energy ΔH required to raise the gold from ambient temperature to an intermediate temperature between its melting point (1336 K) and 3000 K is proposed to be as follows, from other findings reported by Gagliano *et al.*:

$$\Delta H = [0.0355T + 2.69]\text{calg}^{-1} \tag{5.22}$$

where T is the temperature in K.

If this equation is assumed to remain valid to the boiling point of gold (3264 K) and for a heat of vaporization of 446 calg^{-1} (reported data), the energy needed to vaporize one gram of gold is

$$\begin{aligned}\Delta H &= [0.0355(3264) + 2.69 + 446]\\ &= 565\ \text{calg}^{-1}\\ &= 2360\ \text{Jg}^{-1}\end{aligned}$$

The total energy then needed to evaporate the corresponding small volume of gold then is

$$E = \Delta H m = 1.79 \times 10^{-6}\ \text{J}$$

A laser delivering about 120 μJ was considered necessary for this work. Gagliano and his co-workers selected a Q-switched YAG-Nd laser with a 300 ns pulse, capable of a peak power of 500 W (150 μJ per pulse).

5.8 EFFECTS ON WORKPIECE MATERIAL

During laser machining the high temperatures imparted to the workpiece disturb the surface layers to the extent that a heat-affected zone is formed.

Machining process	Tensile strength (MPa)	0.2% yield strength (MPa)	Elongation
Laser	1070	1060	1.3%
Conventional	1130	1070	11.3%

(a)

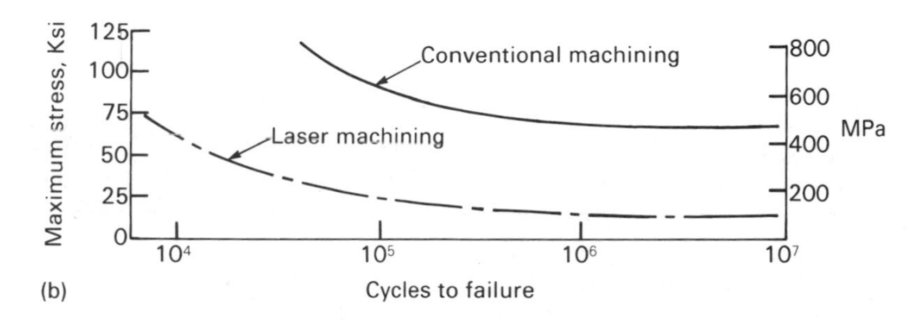

Fig. 5.13 Properties of titanium alloy machined conventionally and by laser (2.25 mm thick). (Contents (%): 6 Al, 6 V, 25 Sn, remainder Ti.) (After Huber and Marx, 1979.)
(a) Tensile properties.
(b) Fatigue characteristics.

The presence of this zone can reduce the fatigue properties of the material. Figure 5.13 shows a comparison of the behaviour of mechanical properties of titanium alloy machined by laser and conventional techniques.

5.9 RATES OF MACHINING AND HEAT-AFFECTED ZONES

Widely varying results have been reported on laser cutting rates and the extent of the zone of the workpiece affected by heat from the laser during machining. A useful review of cutting rates has been presented by Allen, Spalding and Whittle (1975). They point out that the higher reflectivities and thermal conductivities of metals might suggest that high power density lasers should be needed for cutting these materials. Nevertheless, they draw attention to experimental evidence for the existence of a critical thermal threshold, above which a sharp drop occurs in the surface reflectivity. Also they describe the enhancement of machining efficiency by gas-assistance of the laser action (see section 5.10).

Table 5.7 Data on cutting speeds and heat-affected zones (HAZ*) (after Allen, Spalding and Whittle, 1975)

Material	*Thickness (mm)*	*Speed (metre min^{-1})*	*HAZ* or Kerf (mm)*	*Power (kW)*
Paper	Newsprint	>600.00	0.13	0.40
Mylar	0.025	>300.00	0.15	0.30
Quartz	2.0	1.00	0.25	0.40
Fibreglass/epoxy	1.5	3.00	0.25	0.40
Fibreglass/epoxy	12.7	4.6	0.63	20.00
Plywood	25.4	1.5	1.50	8.00
Plywood	17.0	0.5		0.50
Hardened tool steel	3.0	1.7	0.20	0.40
Mild steel	1.2	4.6	0.20	0.40
18–8–1 stainless steel	1.3	4.6	0.20	0.40
Stainless steel	2.5	1.27	0.25	0.40
304 stainless steel	4.7	1.27	2.00	20.00
Titanium	1.0	7.50	0.50	0.60

Typical cutting speeds and the extent of the corresponding heat-affected zones are given in Table 5.7. Huber and Marx (1979) show that the higher the laser cutting feedrate, the smaller is the thickness of the heat-affected zone (Fig. 5.14). They report that for gas-assisted laser cutting, the smaller the diameter of the gas nozzle and the narrower its distance from the workpiece surface the better is the quality of the cut. The gas pressure also plays a significant part in the determination of quality and rate of machining. Typical results are shown in Fig. 5.15 for the cutting of titanium plate. Further improved quality was obtained when the gas pressure was increased from 138 to 276 kPa. In Table 5.8 data are presented on laser cutting of a range of aerospace materials.

Further information on cutting rates for a wide range of materials has been supplied by Taylor (1974) and is given in Table 5.9.

5.10 GAS JET-ASSISTED LASER MACHINING

As noted above, the efficiency of metal machining by laser is often increased by oxygen-assisted gas cutting. This technique is based on the exploitation of exothermic chemical reactions, which are utilized in the well-established oxy-acetylene torch cutting of metals. With the latter effect, the initial melting and oxidation of the metal are caused by the heat from the torch. The cutting is achieved by the release of heat from the oxidation process, and the flow of the gas stream also contributes, by removing the oxide from the cutting area.

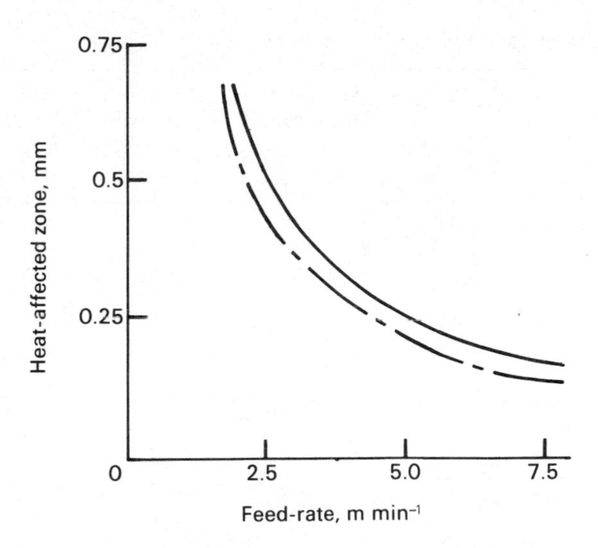

Fig. 5.14 Effect of feed-rate of laser on depth of heat-affected zone of titanium-alloy sheet (1.275 mm thick). (After Huber and Marx, 1979.)

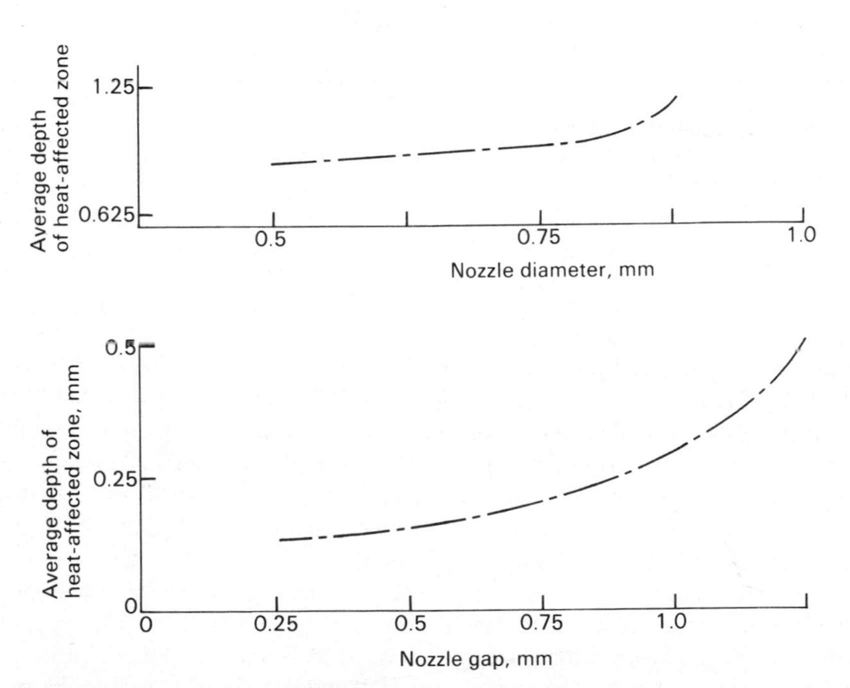

Fig. 5.15 Effect of nozzle diameter and gap width on depth of heat-affected zone. (After Huber and Marx, 1979.)

Table 5.8 Data on laser drilling (after Huber and Marx, 1979)

Material	*Thickness* (*mm*)	*Nozzle diameter* (*mm*)		*Oxygen pressure* (*bar*)	*Max. feed rate for 'best' or 'optimum' quality cut** (*mm* s^{-1})
Ti.6Al.4V.and	0.63	0.66		2.8	127
Ti.6Al.6V.2Sn	1.27	0.66		2.8–4.1	110
	12.70	0.89		4.1–5.5	17
Rene 41	0.51	0.79	0.89	2.8–4.1	34
	1.27	0.89		2.8	13
Hastelloy X	0.51	0.79		2.8	34
	1.27	0.89		2.8	13
Haynes.188	0.51	0.79	0.89	2.8–4.1	42
	1.60	0.79	0.89	2.8–4.1	4
TD.Ni.Cr	0.38	0.79	0.89	4.1–5.5	8
	1.52	0.89		4.1	6
4340 steel	1.60	0.79	0.89	2.8–4.1	25
	3.18	0.79	0.89	4.1	13
410 stainless steel	1.60	0.79	0.89	4.1	25
	3.18	0.79	0.89	4.1–5.5	4

Fixed parameters
Laser output power: 250 watts
Lens: 64 mm focal length (0.127 mm spot diameter at surface)
Nozzle gaps: 0.51 ± 0.25 mm
Assist gas: oxygen
* Straight cuts

For CO_2 laser machining, jet assistance may be discussed in relation to Fig. 5.16. From that diagram the laser output is reflected downwards from a totally reflecting mirror, and is focused through a lens. This lens forms part of a small pressure chamber into which oxygen is fed. Both the focused laser beam and the oxygen jet emerge coaxially through a nozzle at the foot of the pressure chamber. The workpiece is cut as it traverses past this focal point. With this technique, titanium of 0.5 mm thickness has been cut with a CO_2 laser of 135 W at 15 m min^{-1}, the narrow heat-affected zone (or kerf) width being only 0.375 mm. Holes have been bored through 3.175 mm thick boron, and titanium-clad boron, which are otherwise difficult to machine. Similarly, titanium of thickness 10 mm has been cut at 2.4 m min^{-1}, as has copper 0.025 mm thick (with chlorine as the reactive gas).

Table 5.9 Typical cutting speeds standard materials using commercially available CO_2 laser (after Taylor, 1974)

Material	*Thickness (mm)*	*Cutting speed (cm min^{-1})*
Quartz	2	100
Asbestos board (dense)	3	180
Resin bonded fibreglass	3	300
Rubber sheet (dense)	3	500
Rubber sheet (sorbo)	3	1 000
Plywood	18	30
Paper	Newsprint	>60 000
Art board	1	6 000
Formica	1.5	550
Acrylic sheet	1.5	1 500
ABS plastic	2.5	850
Melinex film (mylar)	0.025	>30 000
PTFE	6	100
Titanium	1	750
Stainless steel	1	450
Hardened tool steel	3	170
Mild steel	1.5	450
Leather	5	250

5.11 APPLICATIONS

5.11.1 Drilling

(a) Experimental findings

In early investigations pulsed ruby lasers were used for piercing holes in diamonds (Gagliano *et al.*, 1969). These were to be employed in the fabrication of dies used for drawing 1.4 mm diameter copper wire from 15 to 20–30 gauge for applications involving telephone cables. Two hundred and fifty laser pulses of energy 2–3 J and of duration 0.6ms, at a rate of 1 pulse per s were used to form an opening of 0.46 mm in the die. Laser machining was attractive because there was no physical contact between the laser tool and the material, so that breakage and wear of the drill-bit were eliminated. The location of the hole to be drilled could be precisely located, and large ratios of depth-to-diameter of hole could be obtained.

In the same report by Gagliano *et al.*, the usefulness of the ruby laser in drilling holes in many materials is discussed. Typical depths of metals that can be removed by ruby laser treatment are given in Table 5.10.

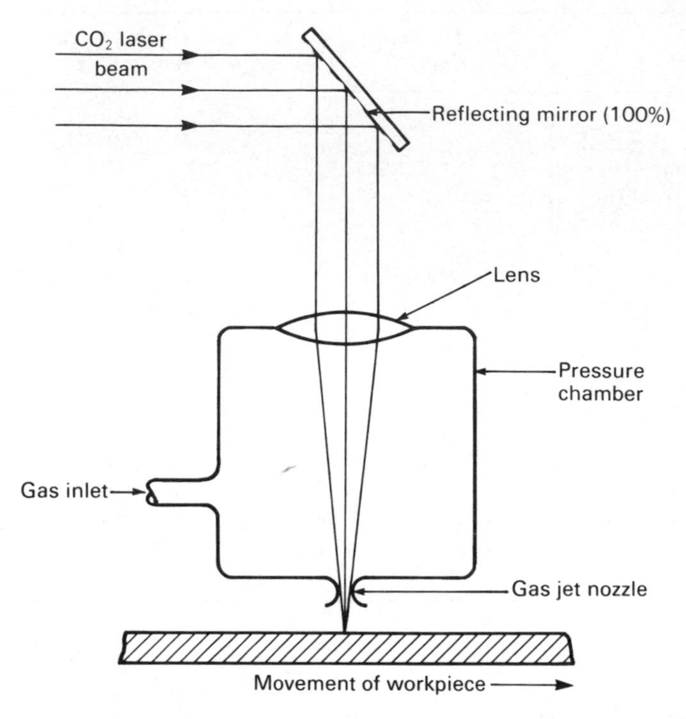

Fig. 5.16 Gas jet-assisted CO_2 laser machining.

Table 5.10 Depth of metal vaporized (after Gagliano *et al.*, 1969)

Material	*Laser energy density* *5 kJcm^{-2}* *600 μs pulse*	*Power density* *10^9 Wcm^{-2}* *44 ns Q-switched pulse*
	Depth vaporized (cm)	*Depth vaporized (μm)*
Aluminium	0.078	3.6
Copper	0.090	2.2
Nickel	0.058	1.2
Brass	0.078	2.5
Stainless steel	0.061	1.1

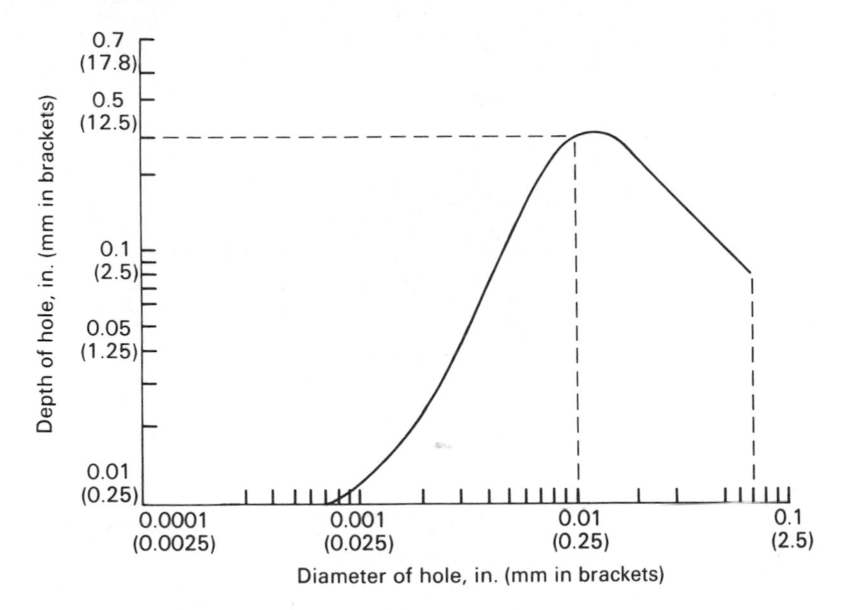

Fig. 5.17 Variation of depth of hole drilled by pulsed laser with diameter. (After Bellows and Kohls, 1982.)

Bellows and Kohls (1982) discuss drilling with multiple pulses, claiming a depth-to-diameter ratio of 40 to 1 on 0.127 mm diameter holes through 12.5 mm thick material. Their Fig. 5.17 shows how the depth of hole produced increases initially as the diameter widens, up to a limit beyond which the penetration decreases. They draw attention to typical small hole applications such as fuel filters, carburettor nozzles, and jet engine blade cooling holes.

The use of laser machining in the aircraft engine industry has also been discussed by Corfe (1983). He describes the steps taken in applying a Nd–YAG laser fitted with CNC control on five independent axes for positioning components weighing up to 1 kg, and of size ranging from 1 m^3 to 100 mm^3 such as gas turbine engine components and nozzle guide vanes. Holes up to 8 mm long and with a depth-to-diameter ratio of 10:1 can be produced with this system, savings of 20 to 30% over EDM being possible.

Boehme (1983) also advises on the use of YAG and CO_2 lasers, pointing out that pulsed solid-state lasers give hole diameters between 0.1 and 0.5 mm, with a drilling capacity of 0.1 to 10 holes per second, the ratio of diameter to depth lying in the range 1:1 to 1:10.

Figure 5.18 shows the section of a hole drilled through a 3.2 mm thick ceramic. In this case, precisely spaced holes, 0.25 to 0.30 mm in diameter in 3.2 mm thick discs of 6.4 mm diameter were required. Their production by

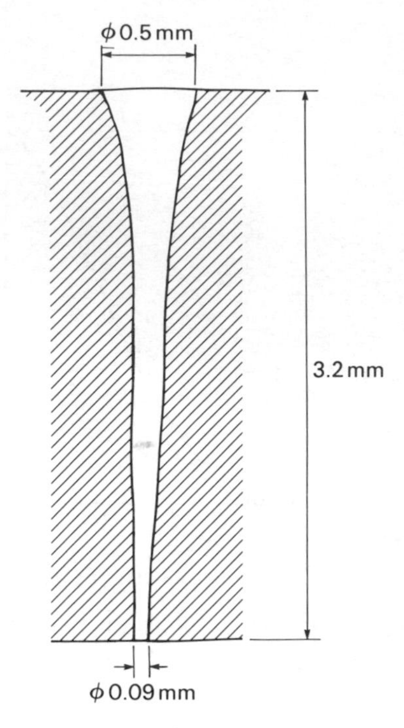

Fig. 5.18 Section of hole in ceramic drilled by ruby laser. (After Gagliano *et al.*, 1969.)

conventional means would have required diamond tipped, hardened steel drill bits. Holes less than 0.25 mm in diameter would have been difficult to achieve, depth-to-diameter ratios of 2:1 being about the limit attainable. Instead the holes were drilled with a ruby laser of energy 1.4 J focused through a 25 mm focal lens onto the surface of the disc. On average, 40 pulses of duration 0.5 ms, at a rate of one per second were used. A 4.75 mm aperture was inserted in the path of the 0.6943 μm laser to maximize the intensity of the beam – such that the energy density was approximately 4 MWcm^{-2} at the surface of the disc – and to refine the formation of the hole.

Figure 5.18 shows the taper produced, from 0.50 mm at the entrance, to 0.09 mm diameter at exit, a depth-to-diameter ratio of 20 to 1 being achieved.

A condition called 'optical piping' which affects the profile of the hole frequently arises in drilling. The effect arises from the internal reflection of laser light as the hole becomes progressively deeper. Normally the laser light is focused at the surface of the hole. The light subsequently diverges on entering the hole, and is reflected from its walls, owing to the small angle of incidence there. The angle of incidence at the foot of the hole is almost normal, and most of the light is absorbed in that region. The maximum hole

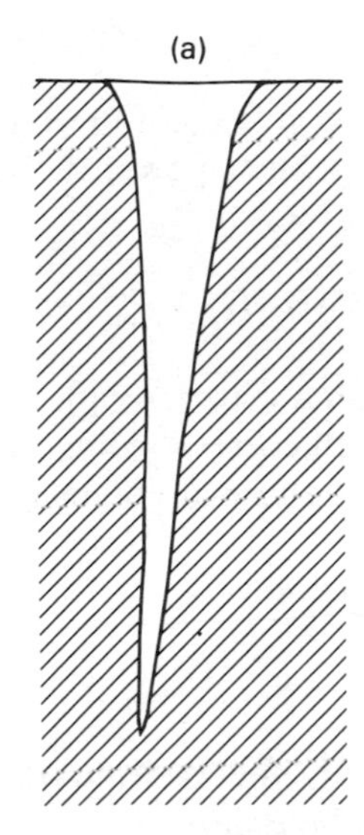

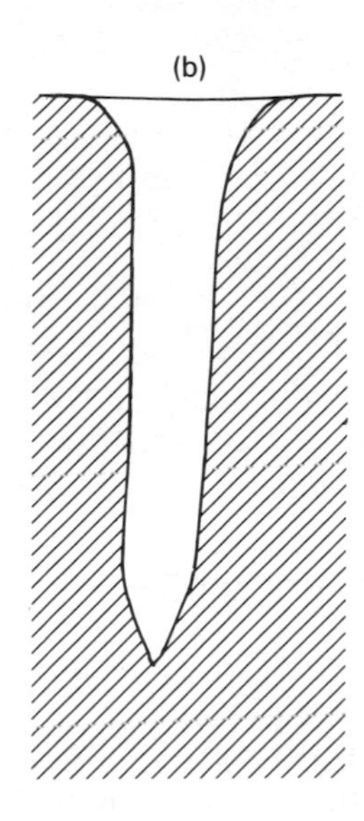

Fig. 5.19 Multiple-shot hole-drilling. (After Gagliano *et al.*, 1969.)
(a) Re-focused.
(b) Not re-focused.

depth achievable is diminished by the energy lost by reflection from the hole wall, and also by decrease in the aperture.

Another effect of optical piping is observed from Fig. 5.19(a) and (b). In this work, attempts were made to re-focus the laser beam after each pulse in order to obtain a deeper hole than would have otherwise been achieved, (since energy is lost by reflection from the hole wall). Figure 5.19(b) (right) shows the profile obtained without re-focusing. Contrary to expectations, as shown in Fig. 5.19(a) (left) on use of re-focusing, a straight hole is not obtained due to 'optical piping'.

Another detrimental effect is the production of a plume of vapour within the cavity, which absorbs the incident laser energy, rather than the metal workpiece. As a result a shallower hole is obtained. To overcome this limitation, and achieve required depths and diameters, repeated pulses are used. Gagliano *et al.* (1969) also propose the use of pulsed lasers operating at reduced energy levels for achieving well-defined holes with reduced taper. With such wide interest in hole drilling by lasers, many attempts have been made to understand the mechanisms involved. Gagliano *et al.* (1969) claim in particular that the generation of a shock wave propagating undirectionally into the material makes a major contribution to material removal.

(b) *Theory*

Similarly, several attempts have been made to give theoretical accounts of hole-drilling. In reports by Weaver, the incident beam power W_1 is expressed as

$$W_1 = 2\pi Dk\{(T_m - T_0)/\ln(L/l)\} \qquad (5.23)$$

where D is the length of hole, drilled through a cylinder with internal and external diameters of respectively L and l, and corresponding internal and external temperatures of T_m and T_0. k is the thermal conductivity of the material.

On the assumption that for steady-state conditions $D = L/2$ and with

$$\ln(L/l) = \ln(2D/l) = 4 \qquad (5.24)$$

Since l appears logarithmically in equation (5.23), there is little variation with this quantity. Then the depth of a long narrow hole, drilled, is

$$D = 4W_1/2\pi k(T_m - T_0) \qquad (5.25)$$

which is approximately independent of the hole diameter.

Example
For a 1 kW laser, used to machine copper where $T_m = 1356$ K, $k = 4.0$ W cm^{-1} K^{-1} the hole depth produced is 0.151 cm.

Table 5.11 shows steady-state hole depths calculated for a range of materials at 300 K temperature. The calculations are based on the assumption that no energy is lost from the incident beam, except that carried away by conduction through the material.

Table 5.11 Calculated depths of holes for a constant incident beam power (after Weaver, 1971)

Material	*Conductivity* ($W\ cm^{-1}\ K^{-1}$)	*Melting temp.* (*K*)	*Depth D (cm)*	
			$W_1 = 10^3$ W	$W_1 = 10^4$ W
Aluminium	2.38	933	0.423	4.23
Chromium	0.87	2176	0.391	3.91
Copper	4.0	1356	0.151	1.51
Gold	3.11	1336	0.198	1.98
Silver	4.18	1234	0.124	1.24
Titanium	0.20	1941	1.64	16.4
Rocks (average)	0.02	2000	15.9	159.0

(c) Moving workpiece

If the component moves at velocity v (cms^{-1}), a part of the beam energy is used in heating the wake of molten material behind the hole. Less beam energy is then available for hole penetration, and less conduction losses (i.e. lower D) are needed for energy balance. The velocity v_c at which the hole penetration is halved is given by

$$v_c = 2\pi\kappa/3l$$

Example
Typically, for metals, thermal diffusicity $\kappa = 1\,cm^2s^{-1}$, and $l = 0.2$ cm. Then $v_c = 10\,cms^{-1}$.

For poor conductors, e.g. concrete (where $\kappa = 0.005\,cm^2s^{-1}$), the velocity is $0.05\,cms^{-1}$.

5.11.2 Cutting

For cutting through steel, powers in the range 200 W to 1 kW are recommended. Oxygen-jet assistance is useful for most metals, as the gas liberates additional energy for exothermal chemical reactions to occur. Bohme (1983) describes the use of a 500 W CO_2 laser in conjunction with a CNC system for cutting steel plates of thickness up to 5 mm.

As shown in Fig. 5.20, from the work of Boehme and Herbrich (1978), with 0.25 kW to 1.0 kW CO_2 lasers the cutting speed for steels decreases on increase in the thickness of the workpiece.

Table 5.8 (from the work of Huber and Marx, 1979) summarizes results from the oxygen-assisted CO_2 laser cutting of a range of steels and alloys of interest to the aircraft engine industry. The growing number of applications for CNC laser cutting of fabrics has been noted by La Rocca (1982). In the production of clothes, a system incorporating a 400 W laser beam, guided by movable mirrors is used to cut a moving fabric, 2 m wide, at a rate of $80\,m\,min^{-1}$, the styles and sizes needed being held in the memory bank of the computer.

The cutting of flat quartz sheet of thickness 5 to 100 mm has been performed by traversing a 100 to 250 W continuous output CO_2N_2He laser focused to a diameter of 25 to 100 μm, across the material. A 300 W laser of the same type has been used to cut through steel plate 3.175 mm thick, at rates of $1.02\,m\,min^{-1}$, the width of the heat-affected zone being about 0.5 mm (Gagliano *et al.*, 1969).

5.11.3 Scribing

Brittle materials like silicon, glass and ceramic may be effectively separated or shaped by laser scribing. With this technique, material is removed by laser

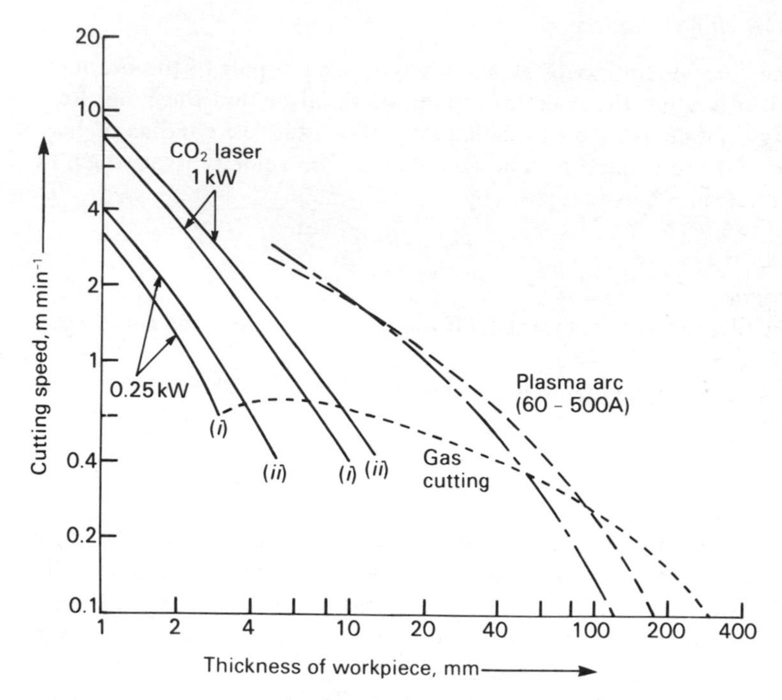

Fig. 5.20 Decrease in speed of cutting with thickness of workpiece. (After Boehme and Herbrich, 1978.)
(i) Austenitic Cr-Ni stainless steel.
(ii) 'St 37' steel.

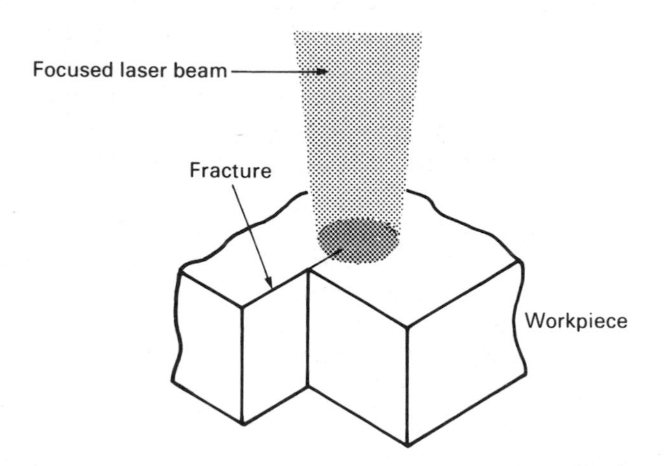

Fig. 5.21 Controlled fracture of brittle material by laser. (After Gagliano *et al.*, 1969.)

along a specific path on the surface of the workpiece. When the material is sufficiently stressed, a fracture occurs along the scribed path.

Silicon transistor wafers have been successfully scribed with a repetitively Q-switched YAG: Nd laser, of peak power 300 W, pulsed at rates of 400 pulse s^{-1} and of duration 300 ns. The beam was focused upon the silicon with a 14 mm focal length lens, which was moved at a rate of 1.5 m min^{-1}.

5.11.4 Controlled fracturing

Laser energy can be used to fracture or break in a controlled fashion delicate items, such as electronic circuits or components. As shown in Fig. 5.21, the beam is focused on a small area of the surface. Its absorption creates thermal gradients which in turn lead to the formation of mechanical stresses. These are sufficient to cause shearing of the material over a region so small that the fracture does not propagate in an uncontrolled manner.

High alumina ceramic, 0.64 mm thick, has been fractured in this way with a 100 W continuous output CO_2–N_2–He laser.

5.11.5 Trimming of electronic components

In a similar fashion, the high power densities, small spot size and short pulse lengths make lasers useful devices for modification of electronic components, often by selective evaporation, and by monitoring, enabling the device to be trimmed or adjusted by a required amount.

In an earlier section, the production of a 10 μm diameter hole in a 0.5 μm thick gold film was discussed. A Q-switched YAG: Nd laser with a 300 ns pulse with peak power 500 W (150 μJ per pulse) was concluded to provide the 1.8×10^6 J energy needed to evaporate the material.

Trimming of resistors composed of chromium-silicon oxide, deposited on the surface of a silicon chip, has been achieved by use of an argon ion laser operated at a power density of nearly 0.8 MWcm^{-2} with pulse widths of less than 10 μs. Energy from the laser was used to heat the resistor film to about 1000°C thereby altering its resistance value.

5.11.6 Dynamic balancing of gyro components

Highly accurate balancing can be also achieved by laser trimming, with material removal at a rate of milligrams per pulse effected by shallow hole-drilling. The dynamic balancing is simplified by the utilization of a signal from the point of imbalance to trigger the laser pulse.

BIBLIOGRAPHY

Allen, T. K., Spalding, I. J. and Whittle, H. R. (1975) The Current Status of Lasers as Industrial Tools, *Proc. IEE Conference on Electrical Methods of Machining, Forming and Coating*, pp. 1–9.

Bellows, G. (1976) *Non-traditional Machining Guide – 26 Newcomers for Production*, Metcut Research Associates Inc, Cincinnati, Ohio, p. 50.

Bellows, G. and Kohls, J. B. (1982) *Drilling Without Drills*, American Machinist Special Report 743, pp. 183–5.

Bloembergen, N. (1979) Fundamentals of Laser-Solid Interaction, American Society for Metals, *Proc. Conf. on Applications of Lasers in Materials Processing*, Washington DC.

Boehme, D. (1983) Perforation Welding and Surface Treatment with Electron and Laser Beam, *Proc. 7th Int. Conf. on Electromachining*, Birmingham, pp. 190–9.

Boehme, D. and Herbrich, H. (1978) Lasers Light the Way to New Technology, *Metalworking Production*, **122** (10), 133–40.

Bolin, S. (1980) Laser Welding Cutting and Drilling, Part 1, *Assembly Engineering*, pp. 30–4.

Carslaw, H. S. and Jaeger, J. C. (1959) *Conduction of Heat in Solids*, 2nd edn, Oxford Univ. Press, Oxford.

Charschan, Sidney S. (1980) *The Evolution of Laser Machining and Welding with Safety*, Ocular Effects of Non-ionising Radiation (Proc. Soc. Photo-optical Instrumentation Engineers), *SPIE*, **229**, 144–53.

Cohen, M. I. (1972) Material Processing, in *Laser Handbook*, vol. 2 (Ed. F. T. Arecchi and E. O. Schulz-Dubois), North-Holland Pub. Co., pp. 1577–647.

Corfe, A. (1983) Why a Laser is Better than EDM for Drilling, *The Production Engineer*, **62** (11), 13–14.

Dunn, J. for Desforges, C. D. (1978) Putting the Laser to Use, *The Production Engineer*, 20–4.

Fowles, G. R. (1975) *Introduction to Modern Optics*, Holt Rinehart and Winston, USA, pp. 264–91.

Gagliano, F. P. *et al.* (1969) Lasers in Industry, *Proc. IEEE*, **57** (2), 114–47.

Harry, J. E. (1974) *Industrial Lasers and their Applications*, McGraw-Hill, London.

Huber, J. and Marx, W. (1979) Production Laser Cutting, American Society for Metals, *Proc. Conf. on Applications of Lasers in Materials Processing*, Washington, DC, 18th–20th April, pp. 273–90.

Kaczmarek, J. (1976) *Principles of Machining by Cutting Abrasion and Erosion*, Peter Peregrinus Ltd, Stevenage, pp. 530–6.

Kato, J. (1985) Investigation of Material Expulsion Mechanism in Laser Drilling Using Modelled Workpiece, *Bull. Jpn Soc. Precision Engineers*, **19** (2), 133–4.

Kock, W. E. (1975) *Engineering Applications of Lasers and Holography*, Plenum Press, New York, pp. 23–30.

Lengyel, B. A. (1966) *Introduction to Laser Physics*, John Wiley and Sons Inc, New York and London, pp. 49–57.

Lepore, M. (1983) Investigation of the Laser Cutting Process with the Aid of a Plane Polarized CO_2 Laser Beam, *Optical Lasers Engineering*, **4** (4), 241–51.

Lunn, D. J. (1981) Nibbling Laser Cutting and Plasma Arc Cutting, *Tooling and Production*, **46** (12), 66–9.
Modest, M. F. (1986) Evaporative Cutting of a Semi-infinite Body with a Moving CW Laser, *J. Heat Transfer, Trans. ASME*, **108** (3), 602–7.
Mori, Masaki and Kumehara Hiroyuki (1976) Study on Ultrasonic Laser Machining, *Annals of the CIRP*, **25** (1), 115–9.
Nichols, K. G. (1967) A Review of Laser Machining, pp. 1–11, Electrical Methods of Machining and Forming, *IEE Conf. Publ.*, number 38.
Rocca, A. V. La (1982) Laser Applications in Manufacturing, *Sci. Am.*, **246** (3).
Rose, C. D. (1985) Ruby Laser Drills Holes with Precision, *Electronics*, **58** (27), 49–50.
Rowe, T. J. and Moule, D. J. (1967) Laser Micromachining of Thin Metallic Films, *Electrical Methods of Machining and Forming IEE Conf. Publ.*, number 61, 19–26.
Schachrai, A. and Castellani Longo, M. (1979) Applications of High Power Lasers in Manufacturing, *Annals of the CIRP*, **28** (2), 457–71.
Schawlow, A. L. and Townes, C. H. (1958) Infrared and Optical Masers, *Physical Review*, **112**, 1940–9.
Schuoecker, D. (1986) Dynamic Phenomena in Laser Cutting and Cut Quality, *Applied Physics B*, **40** (1), 9–14.
Scott, B. F. (1967) Laser Fabrication Processes and Machine Design, Electrical Methods of Machining and Forming, *IEE Conf. Publ.*, number 38, 228–44.
Scott, B. F. and Hodgett, D. L. (1967) Pulsed Laser Machining, Electrical Methods of Machining and Forming, *IEE Conf. Publ.*, number 38, 229–38.
Smith, W. V. and Sorokin, P. P. (1966) *The Laser*, McGraw-Hill, USA, pp. 1–11.
Stovell, J. E. and Scott, B. F. (1970) CO_2 Laser Machining, Electrical Methods of Machining Forming and Coating, *IEE Conf. Publ.*, number 61, pp. 7–13.
Taylor, A. F. D. S. (1974) The Performance of Lasers as Machine Tool Systems, *Proc. 14th Int. Conf. on Machine Tool Design and Research* (Ed. F. Koenigsberger and S. A. Tobias), Macmillan Press, Manchester, pp. 579–602.
Tuck, D. (1967) Micromachining and Welding with Pulsed Ruby Laser Beams, *Proc. IEE Conf. on Electrical Methods of Machining and Forming*, London, pp. 245–51.
Weaver, L. A. (1971) Machining and Welding Applications, in *Laser Applications*, vol. 1, Academic Press, New York and London, pp. 201–38.
Webster, J. M. (1976) The Application of CO_2 Lasers to Manufacturing Processes, *The Production Engineer*, **55** (7), July/August, 373–7.
Whiteman, P. (1965) Laser Machining of Metals, *Proc. Conf. on Machinability*, Publ. by the Iron and Steel Institute, London, 4–6 October, pp. 239–41.
Wright, J. K. (1979) The Development of Pulsed Solid State Lasers for Industrial Applications, in Laser Advances and Applications (Ed. B. S. Wherrett), *Proc. 4th Nat. Q. Electron. Conf.*, J. Wiley & Sons, Chichester/Heriot-Watt Univ., Edinburgh, pp. 261–8.
Yoshioka, S. (1986) Generation of Acoustic Emissions in Ruby Laser Drilling, *Bull. Jpn Soc. Precision Engineers*, **20** (1), 33–8.

6 Electrodischarge machining

6.1 INTRODUCTION

Metal erosion by spark discharges was first observed by Sir Joseph Priestley as early as 1768. More than a hundred years were to elapse before some practical use was made of the effect. Spark discharges became used increasingly for the disintegration of various metals to produce colloidal solutions, and for the removal of broken taps, drills and reamers. Nowadays spark erosion is well known as an effect that occurs with contact breaker points and the sparking plug electrodes in cars: the spark passing between the two electrical contacts causes the removal of a tiny amount of material from each of them.

In 1943 two Russians, B. R. and N. I. Lazarenko, whilst investigating the wear of switch contacts, deduced that spark discharges could be utilized for machining recently developed new metals which were proving to be difficult to shape by established methods. Their observations were a promising route to follow, at a time when their country was desperately short of the right kinds of diamonds needed for machining these materials.

6.2 LAZARENKO RELAXATION (RC) CIRCUIT

The two Lazarenkos realized that the spark energy would have to be harnessed and controlled if the discharges were to be efficiently utilized for machining. Up to the time of their findings, low frequency arc discharges had been used to disintegrate metal, a technique that is still used in the removal of broken taps. Briefly, with the latter method an arc is formed when a voltage is applied between an electrode and a workpiece which are periodically brought into touch and then retracted by the mechanical vibration of one of them. The entire operation is clumsy, the surface finish is poor, there is no reproduction of surface detail, and much of the discharge energy is wasted, being spread over a comparatively large area.

As an alternative to this method, the Lazarenkos deduced that spark discharges of shorter duration and at high repetition rates had the qualities needed for effective machining, and that by submerging the discharge in a liquid dielectric, the energy could be concentrated onto a tiny area.

To that end, they developed a relaxation, or RC circuit, shown in Fig. 6.1. The tool and workpiece are immersed in a dielectric such as paraffin, and are connected to either side of a capacitor, which is charged from a d.c. source. As indicated in Fig. 6.2, the increase in voltage of the capacitor should be larger than the breakdown voltage and hence great enough to create a spark between the tool and workpiece, at the region of least electrical resistance, which usually occurs at the smallest inter-electrode gap. Erosion of metal from both electrodes takes place there. (From Kohlschutter's observations,

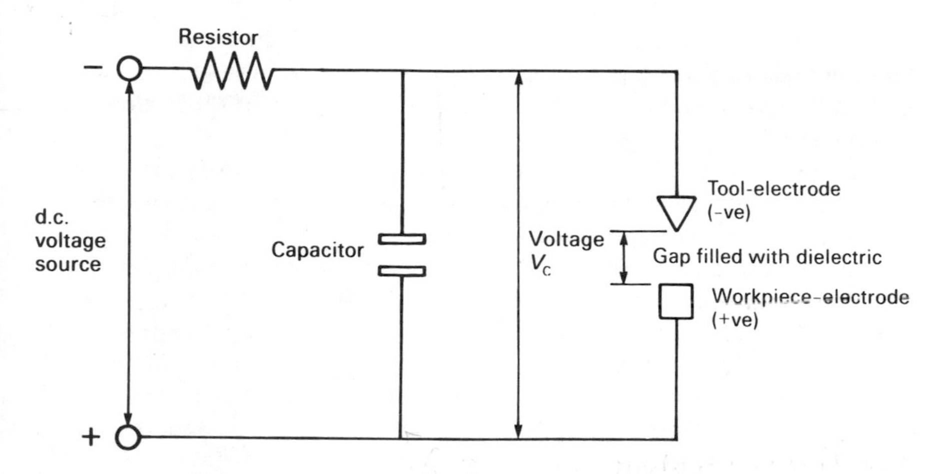

Fig. 6.1 Relaxation circuit.

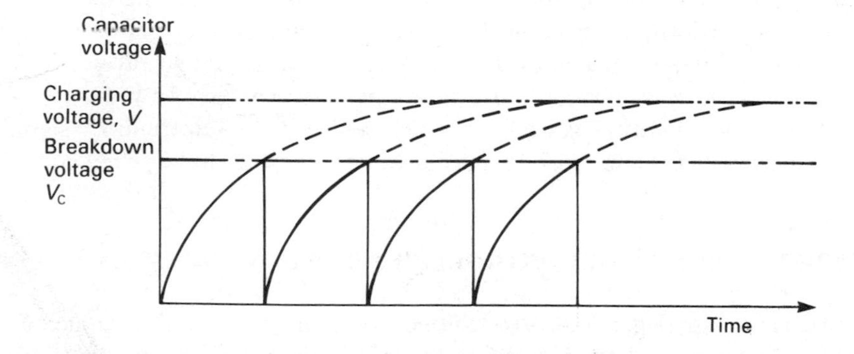

Fig. 6.2 Variation of capacitor voltage with time.

less metal is removed from the negative electrode, which was therefore made the tool in the Lazarenkos' RC circuit.) After each discharge the capacitor is re-charged from the d.c. source, through a resistor (Fig. 6.1), and the spark that follows is transferred to the next narrowest gap.

The dielectric serves to concentrate the discharge energy into a channel of very small cross-sectional area. It also cools the two electrodes, and flushes away the products of machining from the gap. The electrical resistance of the dielectric influences the discharge energy and the time of spark initiation. If the resistance is low, an early discharge will occur. If it is large the capacitor will attain a higher value of charge before the discharge spark occurs. That is, the energy and repetition rate of the sparks depend on conditions in the inter-electrode gap. These conditions in turn are affected by the debris produced by the metal erosion.

The spark temperature is estimated to be as high as 20 000°C. Since its duration is short, and the area over which it is applied is so small, the spark melts and vaporizes a tiny portion of the workpiece, with little effect on the adjacent regions.

The cumulative effect of a succession of sparks spread over the entire workpiece surface leads to its erosion, or machining to a shape which is approximately complementary to that of the tool.

As the workpiece is spark-eroded, the tool has to be advanced through the dielectric towards it. A servo system is needed to ensure that the electrode moves at the proper rate, to maintain the right spark gap, and to retract the electrode if the gap is bridged by particles of swarf. The servo system has therefore to be highly responsive to ensure efficient machining. Systems which compare the gap voltage with a reference value are employed; they are capable of advancing or retracting the electrode.

The electrodischarge machining (EDM) process is not influenced by the hardness of the workpiece metal. To machine a workpiece to a required shape, a tool-electrode of the image form is normally used. The electrode must be electrically conducting; brass was the usual choice for the earlier generators. Modern machines are able to use a wider range of electrode materials. The way towards these developments was paved by the introduction of new kinds of generators which replaced the Lazarenko system, enabling significant improvements in machining efficiency to be achieved.

6.3 DEVELOPMENT OF CONTROLLED PULSE GENERATORS

The early RC spark generators were fitted to converted conventional machine tools. They gave limited satisfaction, although improvements in their performance led to increases in the metal removal rates from about 20 to

$250 \times 10^{-6}\ m^3\ hr^{-1}$, in the period 1950 to 1960. Another drawback was that with RC circuits, finer surface finishes were only obtainable at the expense of metal removal rate: for a finish of $0.3\ \mu m$, the removal rate is only about $0.02 \times 10^{-6}\ m^3\ hr^{-1}$. These comparatively low metal removal rates initially restricted the application of EDM to special applications, in particular die-sinking and the manufacture of press-tools.

Some of the limitations encountered with the RC system may be appreciated from consideration of Fig. 6.2. From the typical variation of condenser voltage with time shown, for the major proportion of time spent on an EDM operation the capacitors are receiving their charge and no effective work is

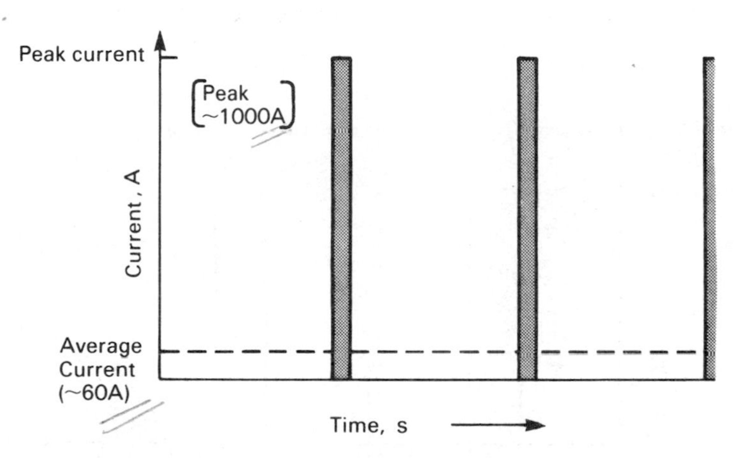

Fig. 6.3 High peak and very rapid decline in current, on spark initiation.

done on the component. From Fig. 6.3, there is a very high peak in current at the instant of spark initiation, followed by a rapid rate of decline. The spark temperature resulting from this high current peak is much higher than that needed to remove a particle of material from the workpiece, and can result in thermal damage of the tool-electrode. That is, the short duration of the discharge and long idle period place considerable restraints on the rate of machining.

Clearly, reduction in peak current values and increase in the spark duration would lead to lowering in the electrode wear, and increase in machining efficiency. These objectives were achieved by the introduction of the controlled pulse generator. Its typical waveform is shown in Fig. 6.4. In comparison with Fig. 6.3 the reduced peak current values, shortened idle periods, and an increase in pulse duration are noted. The ratio of 'on' and 'off' times is called the duty cycle. Since these process variables can be readily adjusted, machining conditions can be selected for particular effects needed.

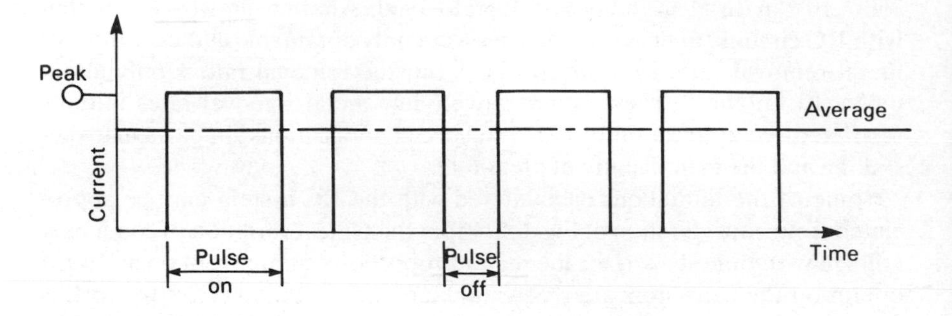

Fig. 6.4 Typical current/time waveform for controlled pulse generator.

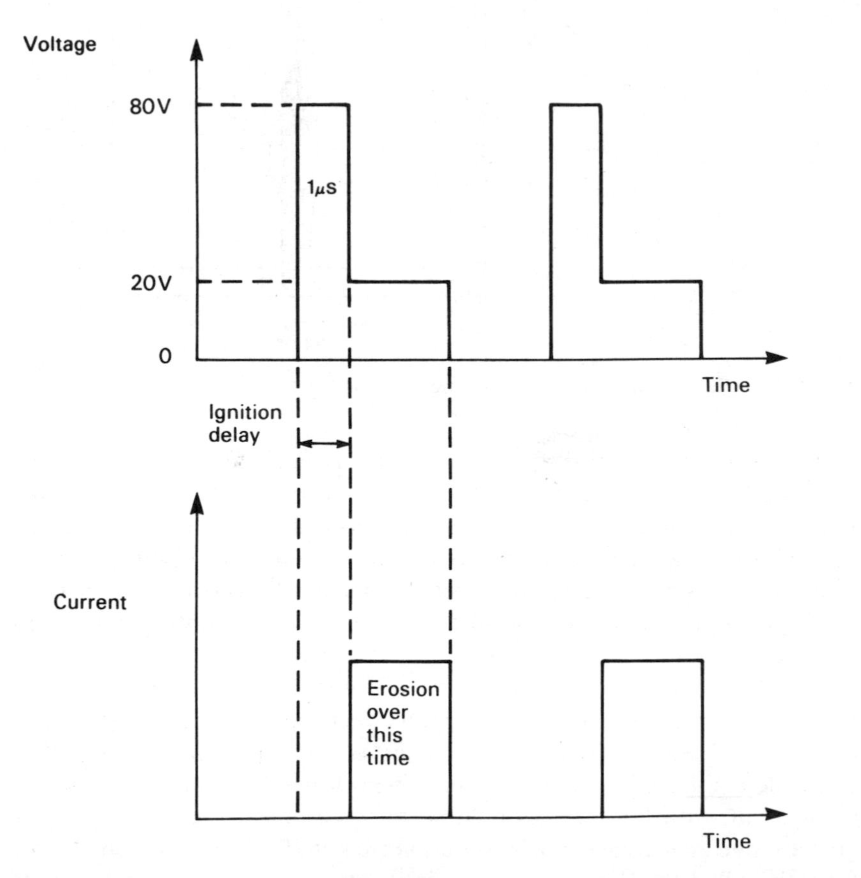

Fig. 6.5 Typical voltage and current characteristics for controlled pulse generator. (Ignition delay is period elapsing before breakdown.)

Thus pulses of high energy and low frequency are used for 'roughing': relatively high material removal with rough surface finish. Pulses of low energy and high frequency yield a finer finish, at the expense of a slower rate of metal removal. A modern electrodischarge machine exhibits typical voltage–current–time characteristics illustrated in Fig. 6.5.

The pulse generator receives its energy from a d.c. source which is fed via a resistor and an electronic switch, to the machining gap. The magnitude of the current, normally 1 to 100 A, is determined by the value of the resistor, the power-supply voltage, usually 60 to 120 V, the duty cycle and the arc voltage. The pulse times are typically 200 μs (on) 0.1 to 10 μs (off).

In comparison with the RC circuit, the pulsed generator enables higher metal removal rates to be obtained with greatly reduced tool wear, for a given surface finish. However, very fine finishes are usually unobtainable with standard transistorized pulse generators. For final finishing, some electrodischarge machines incorporate both pulsed and RC circuitry.

The electrodischarge action occurs primarily at the smallest gap between the two electrodes. This position changes in response to the spark discharges which occur rapidly at different places over the workpiece surface. It soon became apparent in EDM that the gap size had to be controlled, in order to ensure stable conditions and to protect the electrode surfaces against arc damage arising from electrode contact. To that end servo-mechanisms were developed to control the gap width. (It is not measured directly, but can be inferred from the average gap voltage.) D.c. or stepping motors or electrohydraulic systems were employed. These have a high reaction speed in order to respond to short-circuits, which increase the tool-wear, or to open-circuit conditions, which cause spark-time to be lost, and hence reduce the machining-rate.

6.4 MECHANISM OF MATERIAL REMOVAL

At this stage, the principles underlying metal removal in EDM can be usefully summarized.

Two metal electrodes, one being the tool of a pre-determined shape, and the other being the workpiece, are immersed in a dielectric liquid such as paraffin or light oil. A series of voltage pulses, usually of rectangular form, of magnitude about 80 to 120 V (and sometimes higher) and of frequency of the order of 5 kHz, is applied between the two electrodes, which are separated by a small gap, typically 0.01–0.5 mm.

The application of these voltage pulses across such a small gap gives rise to electrical breakdown of the dielectric. The process of breakdown of the dielectric is a localized event: it occurs in a channel of radius approximately

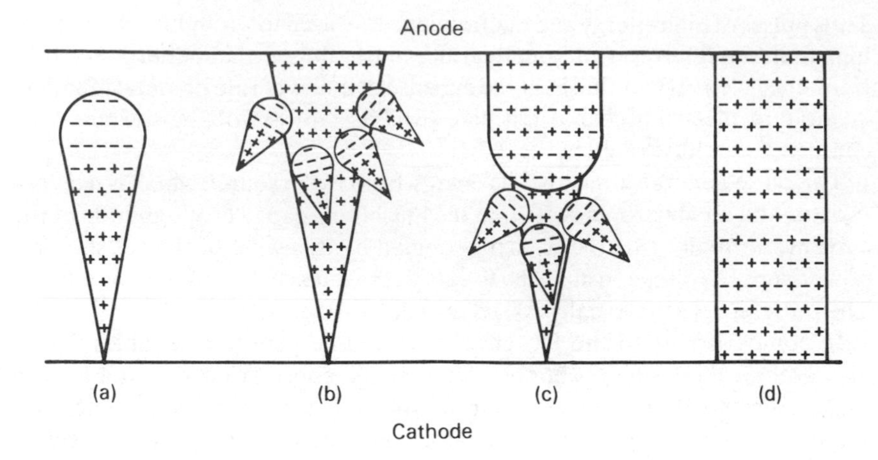

Fig. 6.6 Mechanism of sparking in EDM. (After Meek and Craggs, 1953.)
(a) Avalanche of electrons.
(b) Positively ionized gas in gap.
(c) Secondary avalanches.
(d) Streamer development.

10 μm. As illustrated in Fig. 6.6, the breakdown arises from the acceleration towards the anode of both the electrons emitted from the cathode by the applied field and the stray electrons present in the gap. These electrons collide with neutral atoms of the dielectric, thereby creating positive ions and further electrons, which in turn are accelerated towards, respectively, the cathode and anode.

If the multiplication of electrons by this process is sufficiently high, an avalanche of electrons and positive ions occurs. These eventually reach the electrodes and a current flows.

When the electrons and positive ions reach the anode and cathode they give up their kinetic energy in the form of heat. Temperatures of about 8 000 to 12 000°C and heat fluxes up to 10^{17} $\mathrm{Wm^{-2}}$ can be attained, so that even with sparks of very short time duration (0.1 to 2 000 μs would be typical) the temperature of the electrodes can be raised locally to more than their normal boiling point.

Owing to evaporation of the dielectric the pressure in the plasma channel rises rapidly to values as high as 200 atm. Such great pressures prevent the evaporation of the superheated metal. However, at the end of the pulse, when the voltage is removed, the pressure also drops suddenly, and the superheated metal is evaporated explosively. Metal is thus removed from the electrodes.

The relation between the amount of metal removed from the anode and cathode depends on the respective contributions of the electrons and positive ions to the total current flow. The electron current predominates in the early stages of the discharge, since the positive ions, being roughly 10^4 times more massive than the electrons, are less easily mobilized than the electrons. Consequently the erosion of the anode-workpiece should be greater initially than that of the cathode.

As the EDM action proceeds, the plasma channel increases in width, and the current density across the inter-electrode gap decreases. With the fraction of the current due to the electrons diminishing, the contributions from the positive ions rise, and proportionally more metal is then eroded from the cathode. The erosion of metal from the cathode can be as high as 99.5%, the wear of the anode being kept as low as 0.5%. In EDM, therefore, the cathode-electrode is made the workpiece. The anode becomes the tool, and is shaped in a soft material like graphite or copper to the image of that wanted eventually on the workpiece. The sparks are generated across the inter-electrode gap, generally at regions where the local electric field is highest, each spark eroding a tiny amount of metal from the surfaces of both the tool- and workpiece-electrodes.

The high frequencies at which the voltage pulses are supplied, together with the forward, servo-controlled electronically driven motion of the tool-electrode towards the workpiece, enables sparking to be achieved along the entire length of the electrodes.

By this means, the image shape of the tool-electrode is gradually reproduced on the workpiece, as illustrated in Fig. 6.7. Crookall and co-workers (1973) have derived a theory which describes degeneration of the tool-electrode shape during EDM, their analysis being based on geometrical considerations. The melted material from the electrodes is re-solidified in the dielectric as tiny spheres, their diameters ranging from 2 to 100 μm. This debris is flushed away from the machining zone by the forced flow of the dielectric fluid; a typical velocity of the dielectric would be less than 1 ms^{-1}.

Since the metal removal is effected by the action of electrical discharges, the rate of machining is not influenced by the hardness of the workpiece. The removal rate, which usually ranges from about 0.08 to 25 $cm^3\,hr^{-1}$, rises with increasing current (which normally extends from 0.1 to 500 A) and with decreasing frequency of sparks (typically 50 to 500 kHz). Surface roughness is similarly affected.

Thus in EDM two operations are normally carried out: an initial 'rough' cut to shape the workpiece, performed at high rates of metal removal, and a fine 'finishing' cut, made at low rates of machining. The shaping operation itself depends on the local field intensity being sufficiently high to produce the discharges over the surface of the workpiece; in practice the sparks occur at random enabling components to be machined to close accuracy.

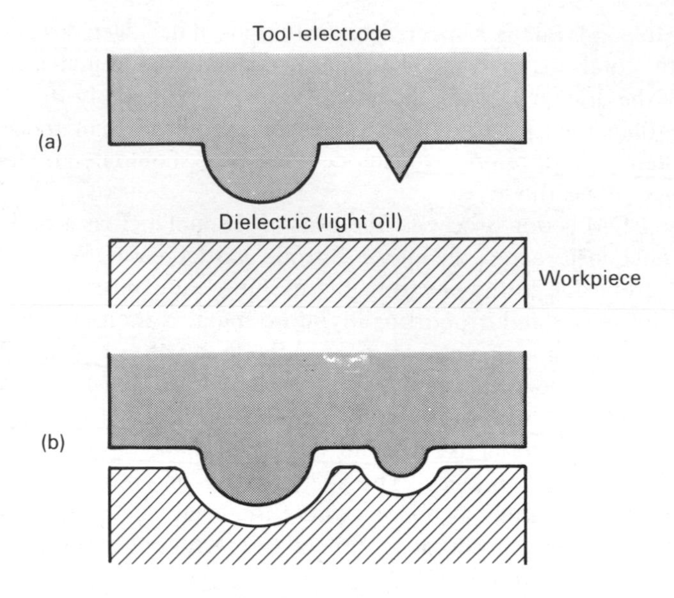

Fig. 6.7 Electrode configuration for EDM.
(a) Initial shapes of electrode and workpiece.
(b) Final complementary shapes of electrode and workpiece.

In brief, therefore, the benefits of EDM are

1. the rate of machining is independent of the hardness of the components;
2. complex shapes can be machined to fine accuracy.

An industrial electrodischarge machine is shown in Fig. 6.8.

Radio frequency signals have recently been found to be a simple and reliable way of monitoring the EDM process and of discriminating between spark and arc discharges (Battacharyya and El-Menshawy, 1978b). Adaptive control systems for EDM machines based on this discovery are now incorporated into industrial equipment (El-Menshawy and Ahmed, 1985).

6.5 DIELECTRIC FLUIDS

The main qualities required for the dielectric fluid are that it should:

1. possess sufficiently high dielectric strength to remain electrically non-conducting until the breakdown voltage is reached.
2. be able to de-ionize rapidly after the discharge.

Fig. 6.8 An industrial electrodischarge machine (left), with its generator and control unit (right). (Courtesy of Transfer Technology Ltd.)

Liquid hydrocarbon products such as paraffin or light transformer oils, and more recently de-ionized water, are mainly used as the dielectric fluids in EDM.

These fluids all possess sufficiently high dielectric strength to remain electrically non-conducting until the breakdown voltage is reached. They are then able to de-ionize rapidly after the discharge has taken place, enabling a high spark-repetition rate to be established. Furthermore their heat capacity is sufficiently high for them to be effective coolants, and being of low viscosity, they are efficient vehicles whereby the machining debris can be flushed out of the spark gap.

The recent introduction of distilled water as the dielectric stems from the benefits obtained in high accuracy EDM. Typical machining conditions are the low energy per pulse (about 10^{-8} J) and pulse durations (typically 0.03 μs), the spark-gap being about 1 μm. Typical frequencies up to 10^5 Hz are used. Under these conditions, surface finishes of the order of 0.06 μm can be achieved.

Although de-ionized water has a much lower conductivity than that of mains water, typically 2 to 5×10^{-16} S m^{-1} compared with 7×10^{-3} S m^{-1} at 25°C, and is therefore very attractive for fine machining, it does suffer the drawback of causing corrosion on ferrous workpieces. Unfortunately, the addition of rust inhibitors serves to increase its electrical conductivity to unacceptable levels.

At the start of EDM, the dielectric is free of eroded particles and other debris. At this stage, its dielectric strength is comparatively high, and an ignition delay occurs before its insulating properties are broken down, and the discharges occur. The debris so produced in the gap reduces the dielectric strength and the discharge action is then able to proceed properly. But if the amount of debris in the gap becomes too great, the particles can form an electrically conducting path between the electrodes, causing unwanted discharges which become arcs, with consequential damage to both tool and workpiece.

The unwanted accumulation of particles in the gap is removed by flushing of the dielectric through the machining zone. Flushing plays a significant part in EDM, particularly when deep and complex shapes have to be produced. Generally, if the dielectric velocity is too low, the gap becomes densely clouded with machining products, and deleterious arcing, which damages

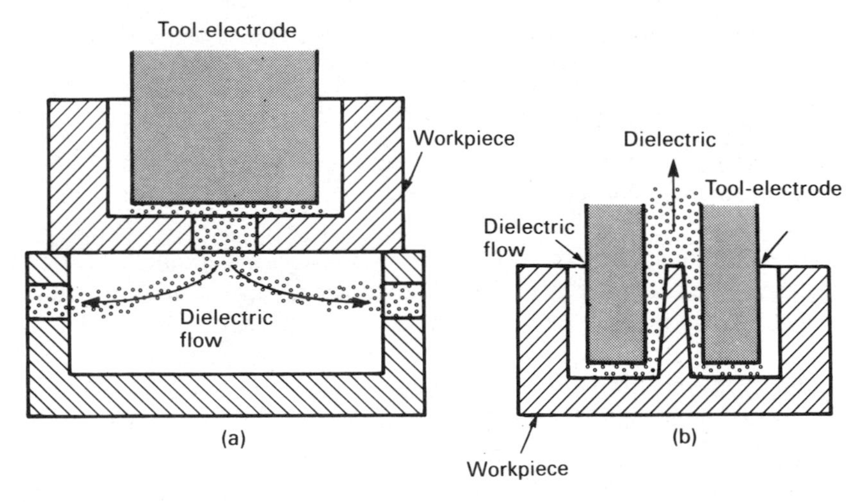

Fig. 6.9 Flushing of dielectric by suction. (After Semon, 1974.)
(a) Through workpiece.
(b) Through tool-electrode.

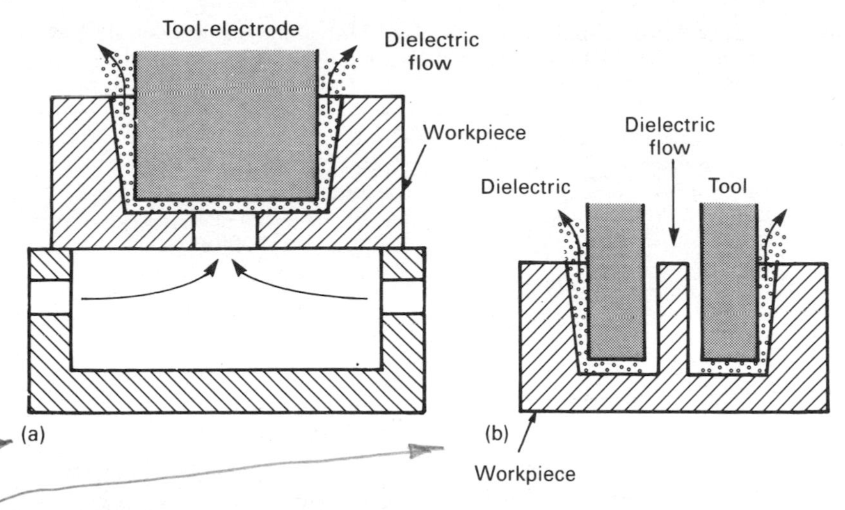

Fig. 6.10 Dielectric flushing by injection. (After Semon, 1974.)
(a) Through tool.
(b) Through workpiece.

both tool and workpiece, can arise. In some cases flushing by suction or by injection is used to give efficient EDM.

As shown in Fig. 6.9 the dielectric may be sucked through either the workpiece or the electrode. This technique is employed to avoid any tapering effect due to sparking between machining debris and the side walls of the electrodes. Injection flushing is another common technique. Two main procedures are illustrated in Fig. 6.10(a) and 6.10(b) with the dielectric being fed through either the workpiece or the electrode which are pre-drilled to accommodate the flow. With the injection method, tapering of components arises due to the lateral discharge action occurring as a result of particles being flushed up the sides of the electrodes. In many modern electro-discharge machines the tool-electrode can be oscillated to improve the efficiency of flushing of machining products from the gap.

6.6 TOOL MATERIALS

Metals with a high melting-point and good electrical conductivity are usually chosen as tool materials for EDM. They should be cheap and readily shaped by conventional methods. Graphite has received much attention as a tool material. Typical information on its main properties, compiled by Hatschek (1984) is summarized below in Table 6.1.

Table 6.1 Properties of graphite electrodes for EDM. Generally low density graphite (1.6 to 1.7×10^3 kg m^{-3}) gives higher removal rates (Data from Hatschek, 1984)

Density kg m^{-3}	*Flexural strength (MPa)*	*Electrical resistivity μohm-m*	*Further information on use for EDM*
1 650	41.4	20.3	Used for EDM of tool steels
1 750	57.2	14.2	Gives good detail, good wear resistance speed, finish
1 840	82.7	12.4	Can be machined to thickness of less than 0.0125 mm. Corner radius to 0.0025 mm
3 390	88.3	1.37	Compacted with copper eliminating 27% porosity
2 970	96.5	1.78	Used for EDM of slots, ribs

A special, high-density graphite is now widely used in pulsed EDM equipment, although the material does not perform satisfactorily in RC EDM work. With the former, it is found to give very low wear, due to its high melting temperature, which is greater than 3000°C. Wear of less than 1% has been reported for rough EDM of steel. This tool material is also capable of yielding high machining rates, a notable exception being obtained with tungsten carbide. The increased use of the material has led to the development of new ways of forming and moulding graphite electrodes.

Copper also has the qualities for high stock metal removal. It is a stable material under sparking conditions. Its wear can be comparable with graphite. Indeed with some workpiece materials, it yields a finer surface finish.

Brass featured as a tool-material in early developments of EDM. Although it is a highly stable material in sparking, its relatively high wear restricts its use to highly specialized applications.

Many other materials have been investigated for use as tool-electrodes in EDM. Cast aluminium electrodes have been used in pulsed EDM. Stock removal with this material is similar to that of copper and graphite but a wear ratio of 15% when used in the rough machining of steel has considerably limited its usefulness. Copper-boron and silver tungsten both exhibit extremely low wear and are occasionally used.

Sometimes copper tungsten is employed as the cathode-metal. Its use yields high machining rates and very low wear. But it is an expensive material, and is not so readily shaped, unlike the other tool materials. Its use is therefore usually limited to applications, such as the cutting of deep holes and slots, for which high accuracy is needed. This tool material is seldom used for rough machining, except in the case of tungsten-carbide workpieces, for which it overcomes the problem of high wear met with other electrode materials.

Computer-aided design and manufacture of tooling for EDM are becoming increasingly attractive, especially for relatively large electrodes. Typical examples include electrodes for large-drop forging dies and car body press tools.

6.7 METAL REMOVAL RATES

No expression has yet been derived that can be used to predict fully the rate of metal removal in electrodischarge machining. This rate depends on a large range of properties of the workpiece material, including its melting point and latent heat. It is also influenced by the properties of the tool-electrodes, and by geometric factors such as the shape and dimensions of the tool and workpiece. Some theories have been developed for analysis of the effects of a single pulse on metal removal rate in EDM. Especially noteworthy is that by Van Dyck (1969) who showed that material erosion can be explained by consideration of thermodynamics and heat transfer models.

Most practical data on machining are available from suppliers of equipment. Typical metal removal rates range from about 0.1 to 10 $mm^3 min^{-1} A^{-1}$. The results given in Fig. 6.11 for the machining of a tool-steel workpiece with a graphite electrode, are representative of the useful information that can be obtained. Not only are metal machining rates quoted, but also effects of machining conditions on surface finish, for a range of currents used in practice.

6.8 SURFACE EFFECTS AND ACCURACY

EDM normally leaves a workpiece with a matt surface, which is covered in small craters of a diameter-to-depth ratio of about 5 to 50. At low powers of EDM, these craters can be about 25×10^{-4} mm in depth, with a diameter of approximately 125×10^{-4} mm. When higher powers are used, the size of the craters can be increased by more than 30 fold, to typical depth and diameters of respectively 12 and 60 μm.

The formation of these craters is a consequence mainly of the discharge action, although it is also affected by the dielectric fluid used and by the

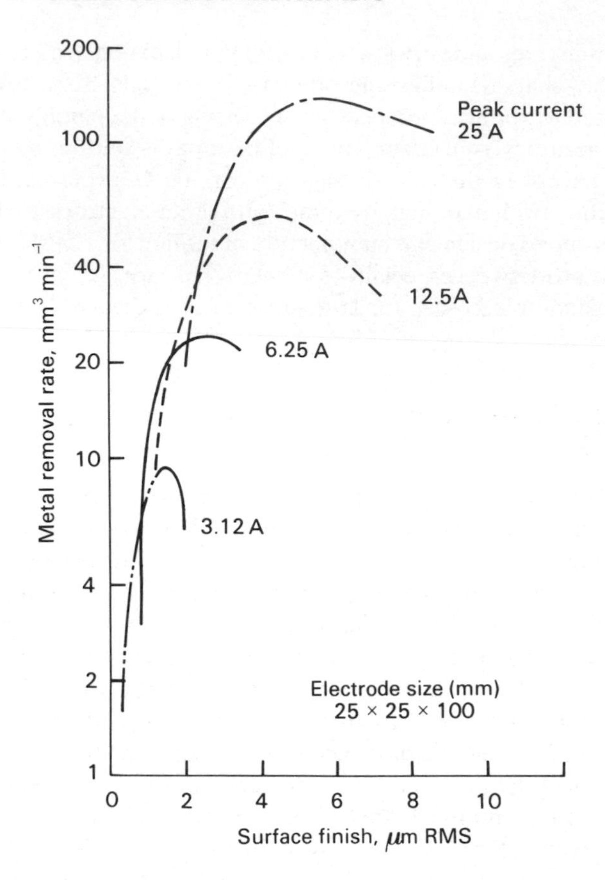

Fig. 6.11 Characteristics of EDM with graphite. (After Hatschek, 1984.)

electrode materials. With the temperatures of the discharges reaching 8 000 to 12 000°C, metallurgical changes occur in the surface layers of the workpiece. Three zones can normally be detected (Fig. 6.12).

The surface of the workpiece is melted and quickly re-hardened by the cooling action of the dielectric so that a thin re-cast epitaxial layer is formed. The layer can be as thin as 1 μm at low powers (e.g. 5 μJ), although it is usually much thicker, by as much as two or more orders of magnitude (e.g. 0.025 mm), at higher powers of EDM.

Adjacent to the re-solidified layer is the heat-affected zone, which is usually less than 0.025 mm in thickness. (A useful rule of thumb is that the thickness of this layer is ten times the surface roughness.) This region is formed by the heating and cooling, and diffusion of material from the melted re-cast layer. The thermal stresses which arise in the melting process can cause severe

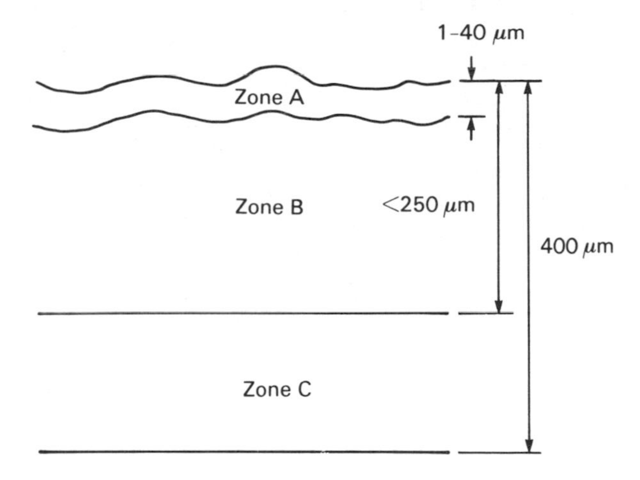

Fig. 6.12 Heat-affected zones in EDM.
Zone A: re-cast layer (1 to 40 μm).
Zone B: heat-affected zone (less than 250 μm).
Zone C: conversion zone (about 400 μm).

metallurgical damage in the heat-affected zone. In particular, grain boundary weaknesses are accentuated and grain boundary cracking is not uncommon. As a result fatigue strength is often reduced.

The removal of the recast layer has been found to increase the fatigue strength by only about 5%. However when both layers are removed, the fatigue strength can be restored to about 95% of its original value. Below the heat-affected zone can be found a 'conversion zone' where a change in grain structure from the original structure is apparent.

The production of these metallurgically damaged surface layers means that fine, conventional finishing is often used on components produced by EDM before they are put into service.

Surface roughness is usually found to decrease with rise in pulse frequency and with reduction in current. Typical surface finishes are about 1.6 to 3.2 μm, although recently claims of 0.05 to 0.1 μm have been reported. Normal tolerances are about ±25 μm with ±5 μm obtainable by judicious choice of process variables.

As indicated on Fig. 6.13 rougher surfaces are produced at longer pulse lengths; the greater the latter process variable, the lower the wear ratio. From the figure it is inferred that if a comparatively rough surface can be accepted, then a much greater amount of workpiece material can be removed before the tool-electrode wears too much or loses accuracy.

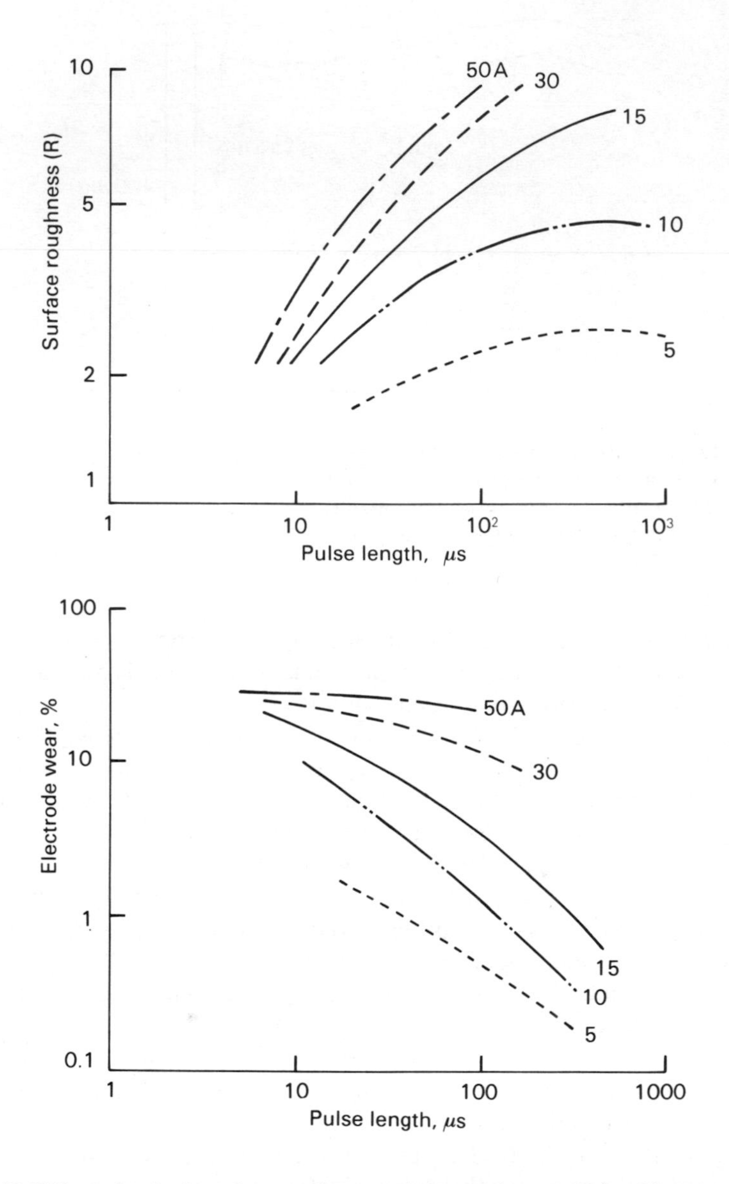

Fig. 6.13 Effect of pulse length on surface roughness and wear. (After Huntress.)

6.9 APPLICATIONS

6.9.1 Drilling

In drilling by EDM, hollow tubular electrodes are commonly used, the dielectric being flushed down the interior of the tube, in order to facilitate the removal of machining debris from the hole. Occasionally solid rods have to be used as the electrode-tools. In this case, the workpiece may be pre-drilled so that the dielectric can be flushed through the machining zone by either suction or injection. Unlike conventional drilling, rotary movement of the tool is unnecessary, so that irregularly shaped holes which can be tapered or recessed or even curved can be produced. It should be noted that injection flushing in EDM-drilling often leaves a tapered hole which widens towards entry. This effect is caused by the debris being flushed between the side-walls of the electrode, making contact with, and causing sparking between, both electrodes. The overcut is thereby increased.

Feed rates of 0.1 mm min^{-1} are typical of drilling by EDM which is applied to the production of holes of diameter 0.1 to 0.5 mm, an overcut of 0.01 to 0.05 mm being normal, see Fig. 6.14(a). In hole drilling the major sparking occurs at the leading edge of the tool. However as debris is washed up the side walls it can cause sparking which leads to overcutting and tapered holes. The effect is not as great as that in ECM. Hence EDM is usually preferred for accurate holes with high length-to-diameter ratios.

The technique has been used mainly in the aircraft engine industry, for instance in the production of cooling channels in turbine blades made of hard metals, like nickel-base alloys. Another application from that industry concerning the EDM drilling of 36 holes in a single operation in jet-engine, annular combustors has been described by Hatschek (1983). The combustors which are made from Hastelloy and HS 188 alloys are tapered double-walled hollow cylinders that are eventually welded together at the end. Each workpiece has 7000 holes of diameter ranging from 0.75 to 2.5 mm through 1 to 6.25 mm thickness of high cobalt alloy. The electrodes used were 400 mm long rods of high-density copper-graphite, centreless-ground to a tolerance of 0.005 mm. To drill up to 36 holes in one operation the power supply used provided 36 channels each of 20 A, the total current rating therefore being 720 A. A computerized NC system enabled holes to be located to within ±0.25 mm. A drilling technique for producing holes at an angle of 45° around a 62.5 mm radius has also been reported. A wear ratio of 15 to 20% was obtained, electrodes having to be plunge-fed a distance of 3 mm to produce a hole 2.5 mm deep. This effect led to circumferential wear at the end of the electrode, producing a 'bullet-shaped nose'. In order to overcome this problem, an automatic electrode-trimming device was used.

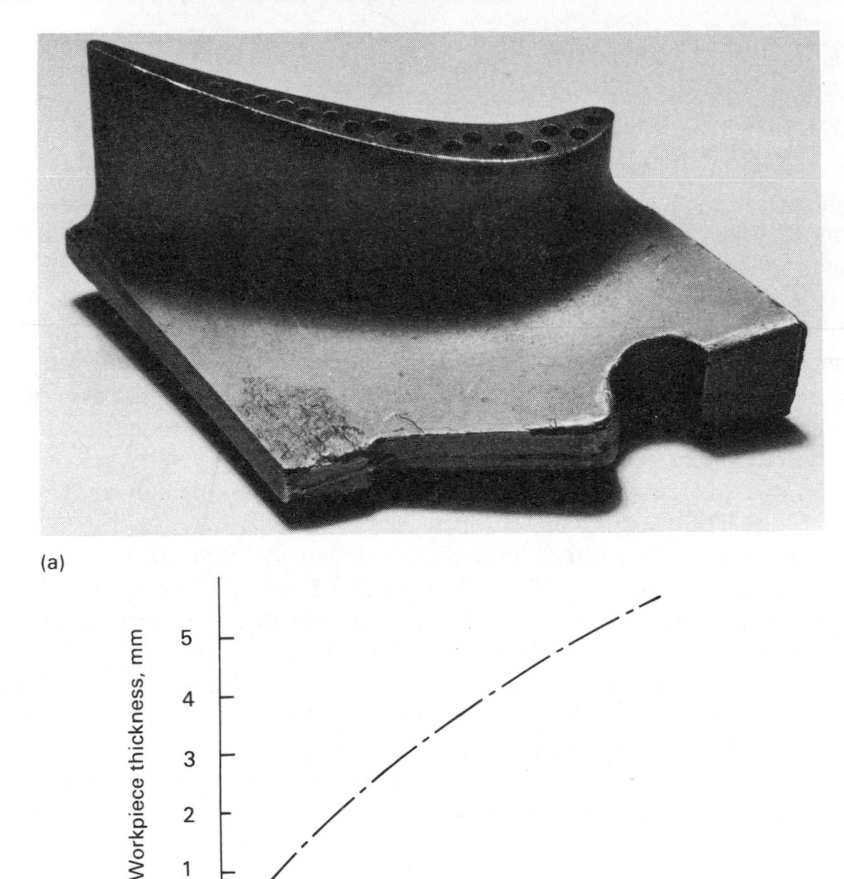

Fig. 6.14 Hole drilling by EDM.
(a) Cooling holes in aerofoil blade.
(b) Dependence of machining time on thickness of component.
(After Toller, 1983.)

Multi small hole drilling has also been discussed by Toller (1983), who draws attention to the increasing number of hole-drilling operations needed for combustors, static vanes and rotating blades. These applications have spurred developments of automatic electrode feed systems, multi-directional electrode guides and multi-headed machines and workpiece transfer mechanisms. As a result, current machines can provide simultaneous drilling with more than 50 electrodes down to 0.3 mm in diameter. For example, Fig.

6.14(b) (Toller, 1983) shows typical times taken to drill 20 holes in nickel-alloy with electrodes of 0.5 mm diameter (based on early trials with tungsten tools).

Marsh (1977) also reports EDM-drilling of 60 holes of 1.27 mm diameter at an angle of 45° in 1.6 mm thick Stellite superalloy jet engine components in 20 min. He also described the drilling of 31 holes from 14.28 mm diameter in components for the car industry (mainly car bumper parts).

6.9.2 Wire-EDM

The use of wire-EDM is a major recent development in the application of the process, with a variety of industrial machines available many of which incorporate computer-numerically controlled systems for guiding the movement of the tool or workpiece and hence the direction of machining. A copper or brass wire, typically of diameter 0.05 to 0.25 mm, acts as the tool electrode and is wound continuously between two spools at rates of up to 3 m min^{-1}. The dielectric, which is usually de-ionized water, is often injected into the machining zone, coaxially with the wire. By this continuous feeding of the wire, a fresh portion of the tool-electrode is always presented to the workpiece which is usually clamped onto a machine table. See Fig. 6.15 and Fig. 6.16. Currents used are typically 2 to 3 A. Hatschek (1984) reviews applications, describing techniques for automatic re-threading of the wire after breakage.

For wire cutting of internal corners with very fine radii, tungsten and molybdenum wires, of diameter as small as 50 μm, are now employed. Other

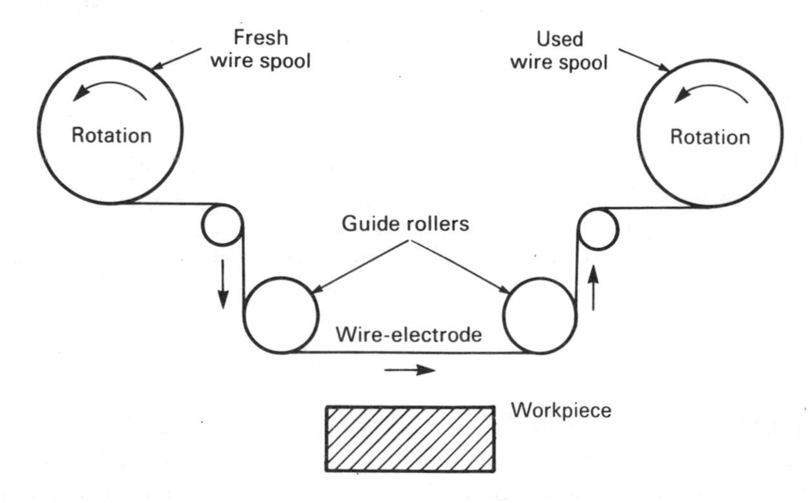

Fig. 6.15 Principles of wire-EDM. (After Marsh, 1977.)

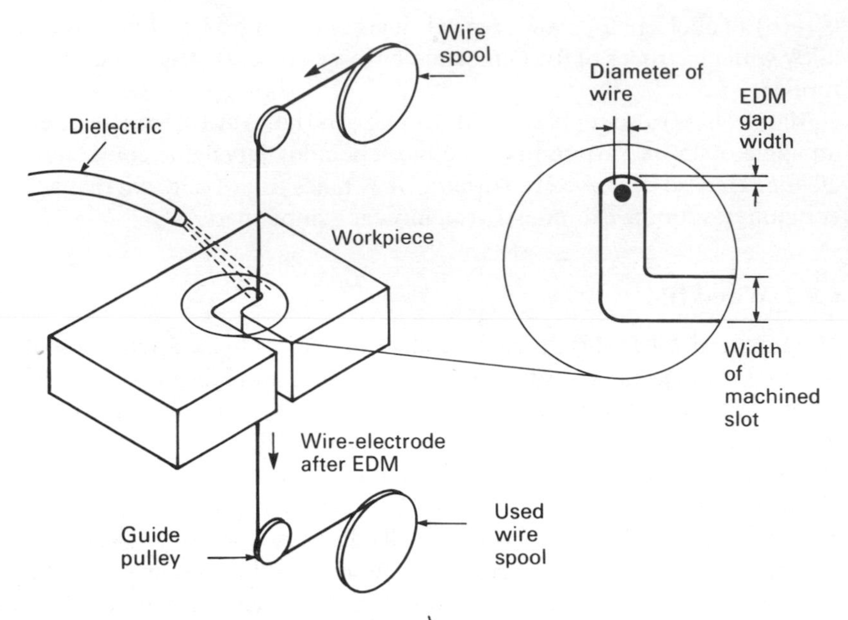

Fig. 6.16 Wire-EDM. (After Bellows and Kohls, 1982.)

wire types include copper coated with a thin layer of zinc, used to raise the current, and hence the energy level of the EDM-discharges, by means of which the cutting speed is increased. Tilting the wire to produce angled cuts has greatly increased the range of applications for this version of EDM.

The wire-electrode technique finds many applications, for instance in the manufacture of press tools, dies and even electrodes themselves for use in other areas of EDM. Huntress describes, for example, how CNC wire-EDM was chosen for the production of rotor- and stator-slot dies for the car industry. He gives details of the difficulties met with the conventional grinding of a rotor blank separating die for the production of components with a 660 mm outside diameter. The tool needed eight kidney-shaped vent holes, and required the die to be made in 32 separate sections, each needing extensive grinding. By wire-EDM the die was produced from only four sections.

6.9.3 Die production

The EDM of three-dimensional items such as dies and moulds is another attractive use of this process. The formation of irregularly shaped apertures in tungsten-carbide extrusion and blanking dies is particularly attractive, although lengthy times of EDM (e.g. 12 hr) are usually necessary.

Forging dies are another example. They are normally made from brass or copper electrodes, which themselves have been produced from a master die.

Crookall and Shaw (1980) have explained some of the reasons for these applications for EDM. With the cavities required the depth is usually less than other orthogonal dimensions, with generous blends and radii which connect adjacent major surfaces. The comparatively low accuracies and modest surface finishes required on the die enable the cavities to be formed quickly, often with only one or two electrodes.

Plastics and rubber moulds are other accurate parts that can be made by EDM, although the finish needed is generally obtained by post-process manual treatment.

A technique that is increasingly employed in EDM applications especially for die production is orbiting of the tool-electrode. This device enables the tool-electrode to sweep a larger or more intricate path than would be obtained from simply sinking the electrode in a straight operation. It enables several different sizes of holes to be machined with one electrode, for instance die holes and clearance holes with one electrode; alternatively a single tool can be used both for roughing, which requires a large gap, and for finishing which needs a smaller gap. Indeed finishing can be faster because the orbiting electrode can produce a series of intermediate steps instead of a single roughing and finishing operation. The final slow finishing operation then removes much less material than would be machined if there were only two steps.

Two major advantages of orbiting are that (*a*) tool wear is spread over the electrode surface, enabling electrode saving of up to 70% to be obtained and (*b*) accuracy is improved. Another benefit is more efficient flushing of the dielectric, owing to agitation caused by the moving electrode. Figure 6.17 shows an example of orbital systems in EDM.

6.9.4 Texturing by EDM

The surface texture of cold reduced steel strips has a significant effect on performance during subsequent processing such as the adherence of paint. Conventional texturing is achieved by shot blasting, which is very difficult to control. Texturing by EDM (EDT) of the surface of the rolls used to roll the strip produces surface textures on the strip to a high degree of consistency and accuracy.

Figure 6.18 shows a roll undergoing EDT. The roll is rotated in a tank where it is covered continuously with a thin film of dielectric. An electrode consisting of a number of graphite or copper segments is also rotated and moved under servo-control to or away from the roll. Repeated discharge at high frequencies causes steady high energy electric discharges which produce

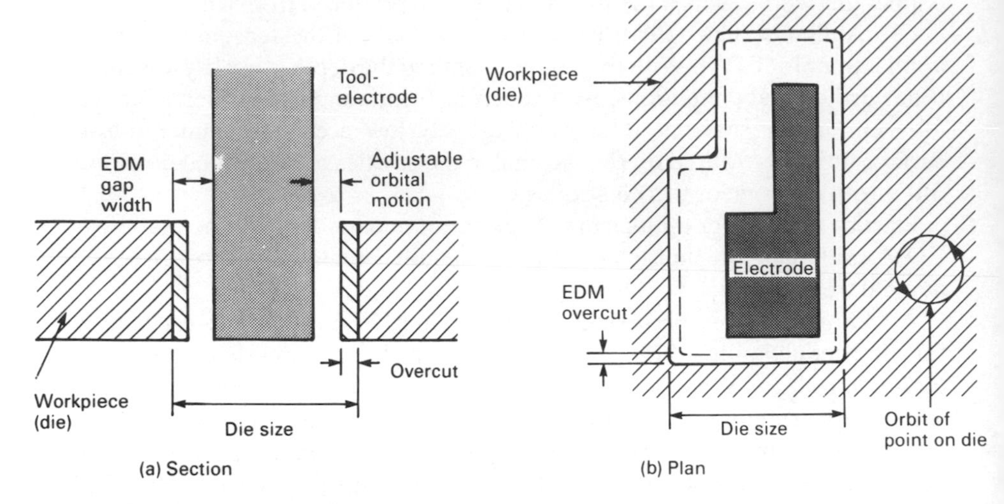

Fig. 6.17 Orbital EDM.
(a) Side view.
(b) Plan view.

Fig. 6.18 A roll being textured by EDM. (Courtesy of Transfer Technology Ltd.)

many minute craters on the roll, resulting in a dense, uniform matt finish, typically 1 to 5 μm *Ra* in roughness (Ahmed, 1987).

BIBLIOGRAPHY

Ahmed, M. S. (1987) Radio Frequency Based Adaptive Control for Electrical Discharge Texturing, *EDM Digest*, **IX** (5), 8–10.

Battacharyya, S. K. and El-Menshawy, M. F. (1978a) *Identification of the Discharge Profile in EDM*, Sixth North American Metalworking Research Conference, April 16–17, University of Florida.

Battacharyya, S. K. and El-Menshawy, M. F. (1978b) Monitoring the EDM Process by Radio Signals, *Int. J. Prod. Res.*, **16** (5), 353–63.

Battacharyya, S. K. and El-Menshawy, M. F. (1980) Monitoring and Controlling the EDM Process, *Trans. ASME, J. Eng. Ind.*, paper 79-WA/PROD-2, pp. 1–6.

Bejar, M. A. (1984) Electro Contact Discharge Grinding of Cutting Tool Metals, *Int. J. Machine Tool Design Research*, **24** (2), 95–103.

Bellows, G. (1976) *Non-traditional Machining Guide, 26 Newcomers for Production*, Metcut Research Associates Inc., Cincinnati, Ohio, pp. 44, 45.

Bellows, G. and Kohls, J. B. (1982) Drilling Without Drills, *American Machinist*, March, 176–8.

Crichton, I. M., McGeough, J. A., Munro, W. and White, C. (1981) Comparative Studies of ECM, EDM, and ECAM, *Precision Engineering*, pp. 155–60.

Crookall, J. R. and Heuvelman, C. J. (1971) Electro-Discharge Machining – the State of the Art, *Annals CIRP*, **20**, 113–20.

Crookall, J. R. and Moncrieff, A. J. R. (1973) A Theory and Evaluation of Tool-Electrode Shape Degeneration in Electrodischarge Machining, *Proc. Inst. Mech. Engrs*, **187** (6), 51–61.

Crookall, J. and Shaw, T. (1980) Why EDM is Expanding from the Toolroom to the Shopfloor, *The Production Engineer*, **59** (1), Jan., 19–23.

De Bruyn, H. E. (1967) Slope Control – A Great Improvement in Spark Erosion, *Annals CIRP*, **14**, 485–9.

El-Menshawy, M. F. and Ahmed, M. S. (1985) *Monitoring and Control of the Electro-Discharge Texturing Process for Steel Cold Mill Work Rolls*, 13th North American Manufacturing Research Conference, May 20–22, pp. 470–5.

Erden, A. and Kaftanoglu, B. (1981) Heat Transfer Modelling of Electric Discharge Machining, *Proc. 21st Int. Machine Tool Des. Res. Conf.*, pp. 351–8.

Gough, P. (1982) EDM Projects Spearhead Government Research Programme into Die and Mould Manufacture, *The Production Engineer*, **61** (5), 33–5.

Guerrero-Alvarez, J. L., Greene, J. E. and von Turkovich, B. F. (1973) Study of the Electro-Erosion Phenomenon of Fe and Zn, *Trans. ASME, J. Eng. Ind.*, No. 11, 965–71.

Hatschek, R. L. (1983) High-volume Hole Making with EDM, *American Machinist*, October, 85–7.

Hatschek, R. L. (1984) EDM Update '84, *American Machinist*, March, 113–24.

Heuvelman, C. J., Horsten, H. J. A. and Veenstra, P. C. (1971) An Introductory

Investigation of the Breakdown Mechanism in Electro-Discharge Machining, *Annals CIRP*, **20** (1), 43–4.

Hockenberry, T. O. (1968) The Role of the Dielectric Fluid in Electrical Discharge Machining, *SAE paper 680635*, Aeronautic and Space Engineering and Manufacturing Meeting, Los Angeles.

Hon, B. (1986) (ed.) Proceedings of Seminar on Electrical Machining, University of Birmingham.

Huntress, E. A. Electrodischarge Machining, *American Machinist*, 207–22.

Jilani, S. T. (1984) Experimental Investigations into the Performance of Water as Dielectric in EDM, *Int. J. Machine Tool Design Research*, **24** (1), 31–43.

Kaczmarek, J. (1976) *Principles of Machining by Cutting, Abrasion and Erosion*, Peter Peregrinus Ltd, Stevenage, pp. 463–86.

Longfellow, J., Wood, J. D. and Palme, R. B. (1968) The Effects of Electrode Material Properties on the Wear Ratio in Spark-Machining, *J. of Institute of Metals*, **96**, February, 43–8.

Marsh, R. (1977) The Changing Face of EDM, *The Production Engineer*, Nov., 18–21.

McGeough, J. A. (1988) EDM/ECM Hybrid Technology, *EDM Digest* (in press).

Meek, J. M. and Craggs, J. D. (1953) *Electrical Breakdown of Gases*, Clarendon Press, Oxford.

Otto, M. Sh. (1983) Effect of the Surface Roughness of the Electrode Tool on Electric Discharge Machining Conditions, *Soviet Eng. Res.*, **3** (11), 97–100.

Rudorff, D. W. (1957) Principles and Applications of Spark Machining, *Proc. I. Mech. E.*, 495–507.

Semon, G. (1974) *A Practical Guide to Electro-discharge Machining*, Ateliers des Charmilles, Geneva.

Snoeys, R. and Van Dyck, F. (1972) Plasma Channel Diameter Growth Affects Stock Removal in EDM, *Annals CIRP*, **21** (1), 39–40.

Somerville, J. M. (1956) *Footprints of a Spark*, Inaugural Lecture, University of New South Wales, Australia.

Toller, D. F. (1983) Multi-small Hole Drilling by EDM, *Electromachining, Proc. 7th Int. Symp.* (Ed. J. R. Crookall), IFS Ltd and North-Holland Publishing Company, pp. 147–56.

Ullmann, W. (1965) Spark Erosion, *Proc. Conf. on Machinability*, London, 4–6 Oct., 1965, The Iron and Steel Institute, pp. 209–14.

Van Dyck, F. (1973) *Physico-Mathematical Analysis of the Electro-Discharge Machining Process*, Ph.D. Thesis, Katholieke Universiteit te Leuven, Belgium.

Vytchikov, Yu. V. and Burda, M. I. (1983) Electric Discharge Machining of the Cemented-carbide Elements of Compression Moulds, *Soviet Eng. Res.*, **3** (11), 100–1.

Wallbank, J. (1980) *The Effect of EDM on Material Properties*, Private Communication and Report.

Willey, P. C. T. (1960) *Spark Erosion Investigations using High Speed Photography and Other Tests*, English Electric Co., Whetstone, UK, Technical Report.

Willey, P. C. T. (1975) EDM Debris Studies Assist Investigation into the Mechanism of the EDM Process in *Electrical Methods of Machining, Forming and Coating*, IEE Conf. Pub. No. 133, 18–20 Nov.

7 Plasma arc machining

7.1 INTRODUCTION

At room temperatures a gas usually consists of molecules, most of which are composed of two or more atoms. (An exception is inert gases, which are not relevant to this discussion.) When the temperature of the gas is raised to about 2000°C, these molecules become dissociated into separate atoms. On further increase in temperature, to approximately 3000°C, some of the atoms have their electrons displaced from them, becoming ionized, that is, electrically charged. A gas in this condition is termed a *plasma*. The presence of a plasma is usually marked by high-frequency sparks.

The prospects for cutting materials by plasma methods were first recognized in the early 1950s when initially the technique was considered as an alternative to oxy-gas flame cutting of stainless steel, aluminium and other non-ferrous metals. The cutting of these metals was virtually intractable by the oxy-gas method, due to the chemical reactions produced. However, even plasma cutting proved to be of limited use. The main reasons were poor quality of machining, the inability of the machines available to handle the high plasma cutting speeds, and the general unreliability of the equipment.

Significant advances have been made since that time as a result of which machining of both metallic and non-conductive materials by plasma methods has become much more attractive. These developments are now discussed.

7.2 BASIC EQUIPMENT

The plasma is commonly produced from a direct current generator, a schematic illustration of which is presented in Fig. 7.1. A hot tungsten cathode and water-cooled copper anode are used to sustain continuously an electric arc between them. A gas is first introduced around the cathode and then flows out through the anode. The smaller the orifice at the cathode, the

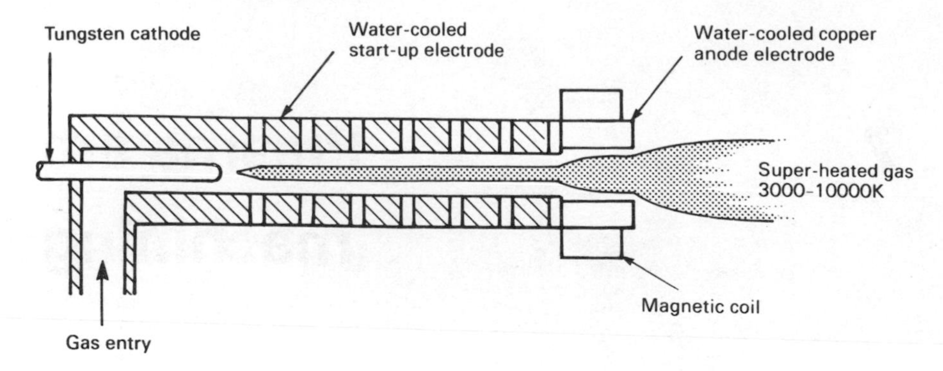

Fig. 7.1 Direct current plasma generator. (From Anon, 1983.)

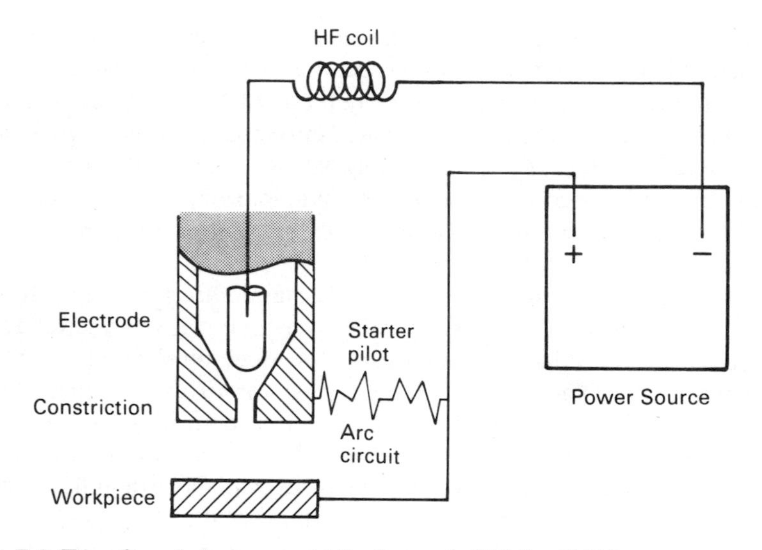

Fig. 7.2 Transferred plasma arc. (After Lucey and Wylie, 1967.)

greater is the temperature – reported to be as high as 28 000° – and hence the heat concentration of the plasma on exit. With such a high temperature when the plasma arc impinges on the workpiece the metal is very rapidly melted and vaporized, the machining debris being removed by the stream of ionized gas.

Plasma devices may be divided into two principal types, which utilize 'transferred' and 'non-transferred' arcs, illustrated in, respectively, Fig. 7.2 and Fig. 7.3. With the former case, the arc is struck from the rear electrode of a plasma torch to the workpiece; the phrase 'plasma arc' is used to describe

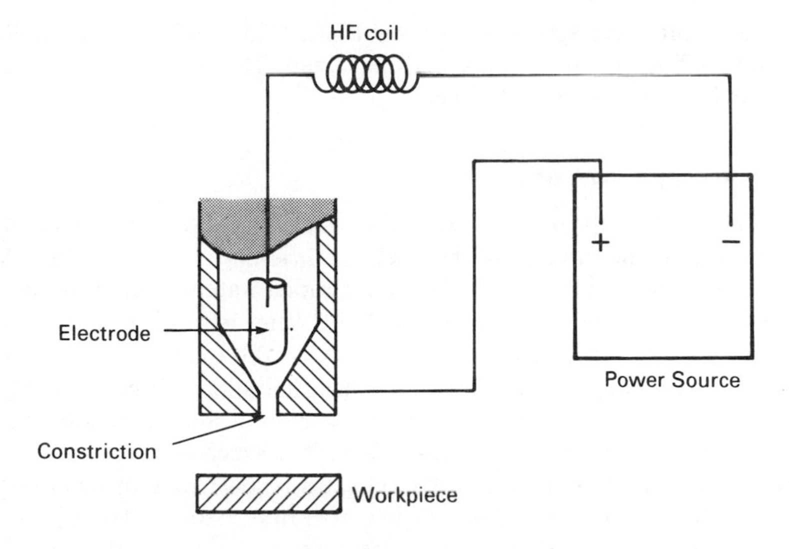

Fig. 7.3 Non-transferred arc. (After Lucey and Wylie, 1967.)

this condition. On the other hand the 'non-transferred arc' is operated within the torch itself. Only ionized gas, or a plasma, is emitted; 'plasma jet' is the term normally employed for this case.

The nozzles first used in plasma machining were found to create two arcs in series, one emanating from the electrode to the nozzle, and the other between the nozzle and the workpiece. This 'double-arcing' effect has been found to damage both electrode and workpiece. Moreover, when several metals have to be machined, for instance, aluminium, stainless and mild steels, of different thickness, different expensive gases (e.g. 65% argon; 35% hydrogen) may have to be employed in order to achieve cuts of acceptable quality. Attempts to overcome this problem have led to the consideration of three alternative plasma systems.

7.2.1 Dual gas plasma torch

Equipment based on this technique has the same general characteristics as the conventional plasma arc devices discussed above. Thus tungsten electrodes are invariably used; the main difference is the addition of an outer shield of gas around the nozzle to reduce the shearing effect of the atmosphere on the cutting gas.

Usually the main cutting gas is nitrogen or argon. Choice of the shield, or secondary, gas depends on the metal being cut. For example, for the machining of stainless steel, aluminium and other non-ferrous metals, hydrogen is often employed as the shield gas. Carbon dioxide gas is also popular with both ferrous and non-ferrous metals. The shield gas may be air or oxygen for a mild steel workpiece.

7.2.2 Water-injected plasma

Figure 7.4 illustrates the main features of this technique. Although nitrogen is still employed as the cutting gas, the shield gas is replaced by water. In order to give maximum constriction of the arc, a radial water-injecting jacket is fitted to the nozzle. This cooling effect of the water is found to reduce the width of the cutting zone. The quality of the cut is also found to be improved because of the cooling effect of the water. Since additional arc constriction is not provided by the water shield, no improvements in the squareness of the cut or in the rate of cutting are obtained through this technique.

In an alternative system (Anon., 1980) the arc is constricted by means of a swirling vortex of water. At the very high temperature associated with plasma arc machining (PAM), estimated in some cases to be about 50 000 K, the water is found to form a film around the arc, instead of evaporating instantaneously as might be expected. That is, the water 'sits' on top of the flame, in a fashion similar to that of mercury on a smooth surface. An insulating

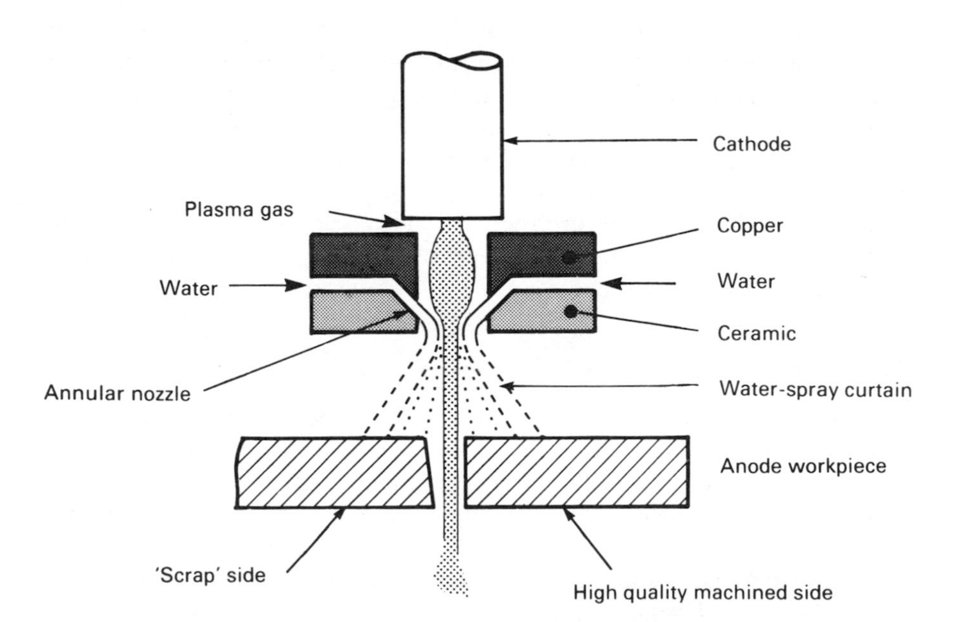

Fig. 7.4 Water-injected plasma system. (After Anon, 1983.)

boundary layer of steam is therefore formed between the plasma and the injected water, thereby preventing further vaporization.

High swirl velocities are, however, found to be necessary for the production of stable water vortices. Moreover the centrifugal force created by the high velocities tends to flatten the annular film of water against the inner bore of the nozzle.

Nevertheless, with this technique only about 10% of the water is vaporized. The remaining amount of water emerges from the nozzle as a conically shaped spray which cools the top surface of the workpiece and inhibits the formation of oxide on the machined surface.

7.2.3 Air plasma

As shown in Fig. 7.5, compressed air is used as the plasma gas. When subjected to the high temperature in the electric arc, the air breaks down into its constituent gases. Since the oxygen in the resulting plasma is very reactive, especially with ferrous metals, cutting speeds are increased, by about 25%.

A disadvantage of this technique is that a heavily oxidized surface is frequently obtained, especially with stainless steel and aluminium. Moreover the air needs to be kept uncontaminated and at the correct pressure. For that purpose compressors have to be used. Hafnium copper or zirconium

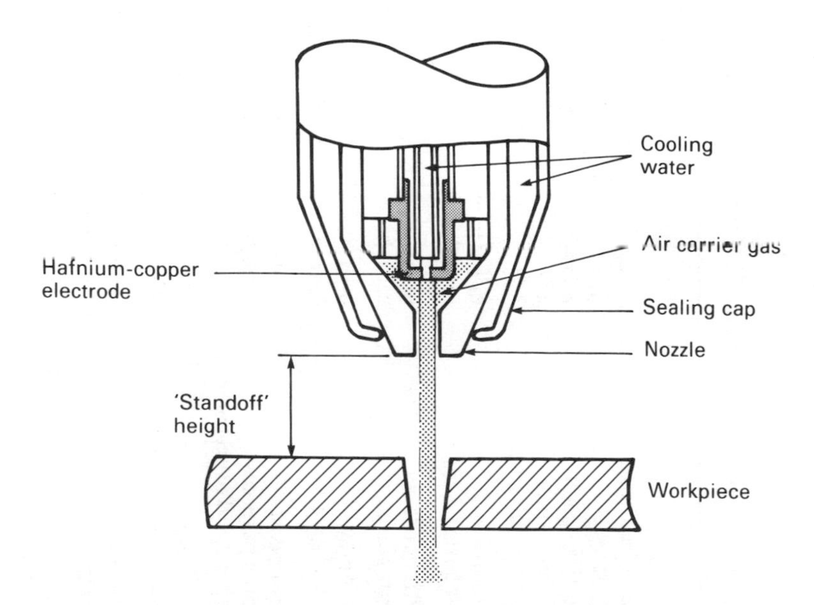

Fig. 7.5 Air plasma system. (After Anon, 1983.)

Table 7.1 Comparison of machining characteristics and costs of plasma arc cutting methods (Data from Anon., 1983)

		Dual gas system			*Water injection plasma system*			*Air plasma system*		
Thickness	mm	3.2	6.4	15.9	3.2	6.4	15.9	3.2	6.4	15.9
Cutting speed	mm s^{-1}	106	51	32	74	64	34	106	95	32
Power setting	kW	15	15	15	58	70	89	49	58	64
Amperage setting		150	150	150	300	350	400	200	235	250
Air flow	m^3h^{-1}	–	–	–	–	–	–	2.8	2.8	2.8
Air cost	cents h^{-1}	–	–	–	–	–	–	0	0	0
Nitrogen flow	m^3h^{-1}	1.7	1.7	1.7	4.7	4.7	4.7	–	–	–
Nitrogen cost	cents h^{-1}	90	90	90	247.5	247.5	247.5	–	–	–
CO_2 flow	m^3h^{-1}	5.9	5.9	5.9	–	–	–	–	–	–
CO_2 cost	cents h^{-1}	525	525	525	–	–	–	–	–	–
Power cost	cents h^{-1}	60	60	60	362	437	556	306	362	400
Torch parts cost	cents h^{-1}	190	190	190	875	875	875	331	331	331
Labour and overhead cost	cents h^{-1}	2000	2000	2000	2000	2000	2000	2000	2000	2000
Total cutting cost	cents h^{-1}	2865	2865	2865	3484.5	3559.5	3678.5	2637	2693	2731
Linear cutting rate	mh^{-1}	381	183	114	267	229	107	381	343	114
Linear cutting cost	cents m^{-1}	7.5	15.6	25.1	13.1	15.5	34.4	6.9	7.9	24.0

electrodes are also employed, in place of tungsten, since the latter material is particularly reactive with oxygen. Irrespective of the materials used, the life of the electrodes is found to be short. Attempts have been made to increase the electrode life by the introduction of oxygen downstream in the nozzle bore through which nitrogen is passed as the main cutting gas. When a gas mixture of 80% nitrogen and 20% oxygen is used the cutting rate of mild steel is found to be increased by about 25%.

Only electrically conducting materials, mainly stainless and chrome nickel alloy steels, aluminium and copper, can be tackled by the air-plasma method. From Table 7.1, air-plasma machining is noted to cost about half that of dual gas and water injection techniques for cutting 6.25 mm thick mild steel plate, due mainly to the use of air as the plasma carrier and shielding gas. Industrial machines carry automatic arc starting equipment which ensures a safe and high initial cutting velocity which can be three to five times higher than that of conventional gas cutting.

7.3 TEMPERATURE EFFECTS

Typical temperatures associated with plasma arc devices are summarized in Table 7.2. They are noted to be considerably higher than those of other methods. Very high maximum heat transfer rates are found to occur with plasma arcs. This effect may be attributed to the transfer of all the anode heat to the workpiece. With the plasma jet (non-transferred arc), since the torch nozzle itself is the anode, a large part of the anode heat is extracted by the cooling water and is not effectively used.

Owing to its greater efficiency the plasma arc system is often preferred for machining metals. On the other hand, non-conductive materials that are

Table 7.2 Maximum temperature heat-transfer rates of plasma arc and comparable devices (Data from Lucey and Wylie, 1967)

	Maximum temperature °C	*Maximum heat transfer rate* MWm^{-2}
Air–propane bunsen flame	1 800	1.64
Air–propane rocket flame	1 800	8.18
Oxy-propane bunsen flame	2 900	8.18
Oxy-propane rocket flame	2 900	19.6
Plasma flame (non-transferred arc)	16 600	68.7
Plasma flame (transferred arc)	33 300	24.5

difficult to machine by established methods can often be successfully tackled by plasma jet cutting.

7.4 MECHANISMS OF PLASMA MACHINING

One of the few theoretical appreciations of material removal by plasma methods is that due to Moss and Sheward (1970). They make the simplifying assumption that the workpiece acts as a heat sink, absorbing all the energy released by the plasma jet incident upon it. That is, the entire heat energy of the plasma jet is available for cutting. (In practice energy is lost by conduction and radiation losses from the workpiece, and by spreading of the jet.) Heat transfer from the plasma to the workpiece is considered to occur mainly by convection, with radiation from the arc column being found to make only a small contribution.

When, for example, an argon plasma is employed, the convective heat transfer is achieved by the recombination of ions with electrons; with a plasma composed of diatomic gases, energy is released by the recombination of atoms. The latter is very effective, probably because it takes place at a lower temperature than the former, and hence nearer to the workpiece.

Moss and Sheward discuss the ways in which material may be removed in plasma machining. Firstly the temperature of the component has of course to be raised to that needed for the material removal reaction to occur. The workpiece may melt, in which case the molten liquid is blown away by the plasma as a fine spray. Flow lines on the machined surface are a common sign of this kind of material removal. A mobile effluent may also be formed by a chemical reaction between the plasma and the workpiece. Another mechanism is vaporization: the vaporized material is removed from the machining zone by the plasma jet; graphite is known to act in this way. When organic-based materials undergo PAM, they pyrolyse forming volatile products which are blown away leaving a solid residue, mainly carbon, which subsequently disintegrates from the surface; alternatively, if the residue is more tenacious it may need removal by other means, such as vaporization.

7.5 THEORETICAL CONSIDERATIONS

Moss and Sheward give a simplified theoretical relationship between cutting speed and arc power dissipation. The rate of energy transfer E from the plasma jet to the workpiece is obtained from the difference in energy of the incident plasma and the effluent, together with the plasma flow rate V m^3/s.

The incident plasma jet at a temperature of T_1(K) is assumed to provide at a rate of energy h_1 (Js^{-1}). The effluent, which is at a temperature T_2(K), is taken to have a heat content of H_2 (Jm^{-3}).

Thus

$$E = Kh_1 - H_2V \tag{7.1}$$

where K is a constant (<1) to account for spreading losses. Also

$$h_1 = eP$$

where e is the torch efficiency, and P is the electrical power (W), supplied to the torch.

Then

$$E = KeP - H_2V \tag{7.2}$$

Now machining is taken to remove a thin rectangular slice of the workpiece.

Thus

$$E = ctv\rho L$$

where c is the kerf width, t is the thickness, v is the cutting speed, ρ is the bulk density (kgm^{-3}). L is the total energy required for the cutting process to convert a unit mass of workpiece material to effluent (Jkg^{-1}).

With equation (7.2),

$$P = (ct\rho Lv + Q)/Ke \tag{7.3}$$

where

$$Q = H_2V$$

Cutting speed is noted to rise linearly with arc power dissipated in the plasma torch, from equation (7.3).

Moss and Sheward have compared their theory with experimental findings for textile materials. As shown in Table 7.3, they used two cutting gases, argon flowing at $2.4 \times 10^{-4}\ m^3s^{-1}$ at approximately 1.4 kW and a 90% argon −10% hydrogen mixture flowing at approximately 2.4 to $2.5 \times 10^{-4}\ m^3s^{-1}$, the powers being, respectively, 1.4 kW and 1.8 kW; the corresponding efficiencies of the plasma torch were 0.65 and 0.75.

Figure 7.6 shows that the torch power increases with cutting speed, the actual amount varying significantly with the mechanism of the removal process.

Table 7.3 Characteristics of cutting of textiles by PAM (Data from Moss and Sheward, 1970)

Material	*Thickness* 10^{-4} *m*	*Density* *kgm*$^{-3}$	*Argon*			*90% Argon – 10% hydrogen*		
			Maximum cutting speed V_M *ms*$^{-1}$	*Limiting cutting speed* V_L *ms*$^{-1}$	*Ratio* r V_M/V_L	*Maximum cutting speed* V_M *ms*$^{-1}$	*Limiting cutting speed* V_L *ms*$^{-1}$	*Ratio* r V_M/V_L
Group 1								
Polypropylene	1.2	618	9.3	0.92	10.1	14	1.15	12.1
Saran	7.2	521		0.33			0.53	
Nylon 6	2	612	14.6	0.63	23	14.6	0.77	19
Nylon 66	2	603	15.8	0.63	25	20.3	0.68	30
Sec cellulose acetate	1.6	650		0.77			0.82	
Polyester	1.6	695	1.1				1.3	
Cellulose triacetate	1.4	655		0.87			0.962	
Glass°	1.0	967	4.9	1.15	4.3	7.8	1.30	6.0
Group 2								
Wool	3	409	1.07	0.38	2.8	1.7	0.48	3.5
Courtelle	6	406		0.53			0.62	
Acrilan	9	240		0.67			0.72	
Group 3								
Cotton	2	673	1.48	0.82	1.8	2.2	0.96	2.2
Flax	2	867	1.06	0.29	3.7	1.51	0.38	4.0
Rayon	3.5	454	1.19	0.38	3.1	1.66	0.48	3.5
Teklan	4.5	444		0.91			1.15	
Dynel	4.9	461		0.72			0.82	
Board	4.1	753	0.02	0.05	0.4	0.02	0.10	0.4
Carbon cloth	2.6	344	0.04	0.10	0.4	0.07	0.20	0.35

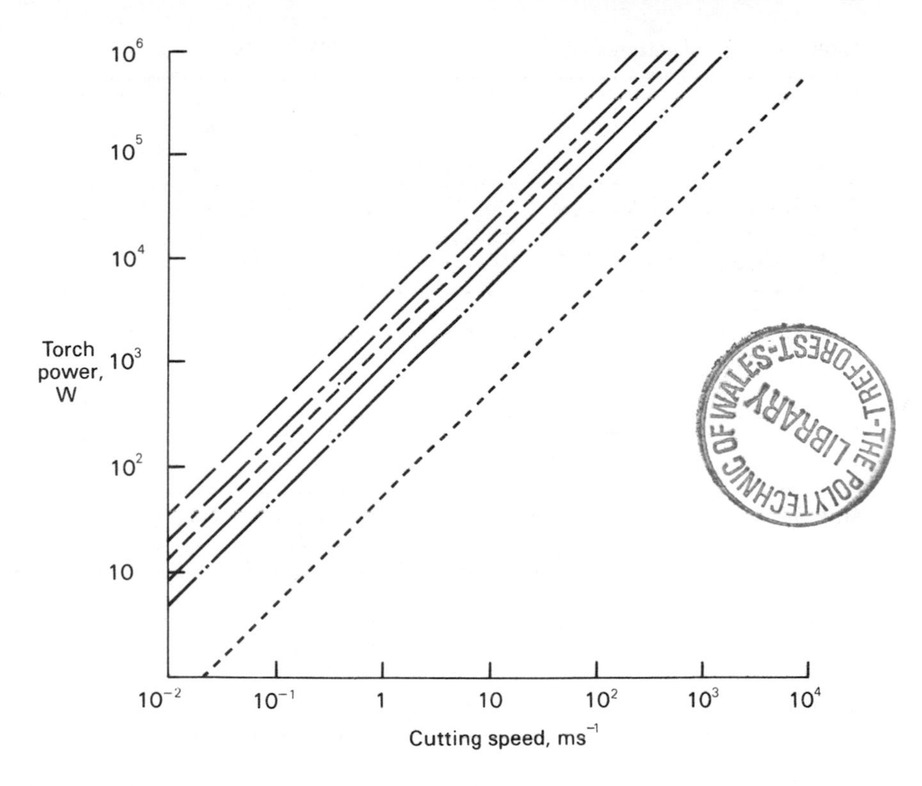

Fig. 7.6 Relationship between torch power and cutting speed. (After Moss and Sheward, 1970.) — — pyrolysis and vaporization (Ar), — - — N_2 reaction (heterogeneous), · — — — — H_2 reaction (heterogeneous), ——— H_2 reaction (homogeneous), —··— pyrolysis, - - - - - - melting.

7.6 REMOVAL RATES

Further information on the cutting of various metals is given in Table 7.4. A useful comparison of plasma arc machining with traditional techniques by Lucey and Wylie, is presented in Table 7.5, showing that its removal rates are substantially higher than those of conventional single-point turning. When compared with oxy-gas cutting at rates of 500 mm min^{-1} through 12.0 mm thick steel plate, rates of 2.5 m min^{-1} can be achieved with plasma machining, although input powers of 220 kW are required (Anon., 1980). Amongst the main process conditions affecting removal rate is surface speed: as this variable is increased, the removal rate is found to rise firstly to a maximum, and then to fall, as indicated in Fig. 7.7. Cutting speed is also found to decrease with increasing thickness of the material, Fig. 7.8. It might be noted

Table 7.4 Typical cutting data (Data from Lucey and Wylie, 1967)

Material	*Arc power kW*	*Argon – 15% H_2 flow rate $m^3\ h^{-1}$*	*Torch angle*	*Cutting time s*
89 mm round bar stainless steel	150	5.7	75° trail	15 s
92 mm round bar En3A	150	5.7	75° trail	37 s
114 mm square 2.25% Cr 1% Mo	150	5.7	90°	1 min

Table 7.5 Comparison of removal rates for conventional methods and PAM (Data from Lucey and Wylie, 1967)

Material	*Conventional turning $mm^3\ s^{-1}$*	*Plasma arc*	
		Roughing cut	*Smooth cut*
Rene 41	128	1365	546
Precipitation-hardening stainless steel	490	1230	546
Inconel	560	1090	410

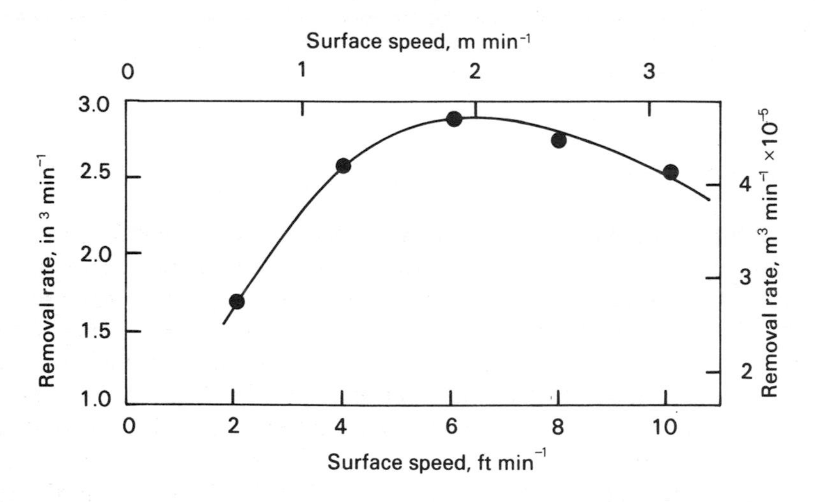

Fig. 7.7 Effect of surface speed on removal rate. (After Lucey and Wylie, 1967.) Power: 32 kW; workpiece: alloy steel.

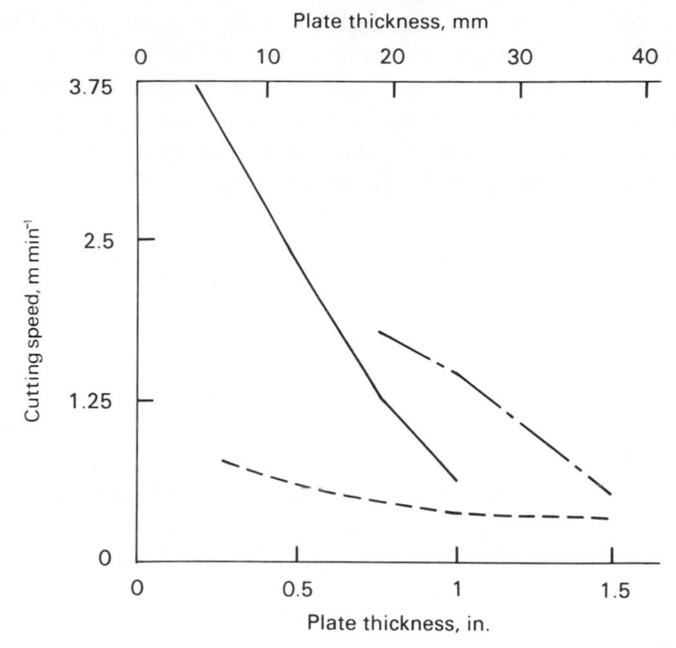

Fig. 7.8 Decrease in cutting speed with increase in thickness of workpiece (mild steel). (From Anon, 1980.) ——— water injection, — - —plasma, — — — oxy–acetylene.

that in bevelling operations by PAM, the cutting speed is determined by the width of the cut face and not the plate thickness. Thus in the case where a 45° bevel is to be produced with a plate of 12 mm thickness, the cut should be set for that of an 18 mm plate, since that is the effective width of face.

As the power is increased, the efficient removal of melted metal during PAM is found to need a corresponding rise in gas flow-rate.

7.7 SURFACE EFFECTS

Bevelling of the edge of the workpiece often arises in plasma machining. Typically one cut edge is square to within ±3°; it is always obtained on the right side of the plasma arc relative to the direction of travel. In contrast the opposite left-hand edge is bevelled to about 15°, this effect being caused by clockwise swirling of the cutting gas.

Most material melted by the plasma flame is removed by its high-velocity gas jet, so that a clean, smooth surface is left. However, owing to the high rate of heat transfer, a depth of fused metal extending to about 0.15 mm below the surface is obtained. Beyond this layer, a further heat-affected zone, of thickness 0.25 to 1.25 mm, can be found (Lucey and Wylie, 1967).

Rapid cooling can cause the propagation of fine cracks beyond the heat-affected zone, often to a distance of 1.6 mm (Lucey and Wylie, 1967).

PAM is not an accurate process. When tolerances better than 1.6 mm are required, final finishing by established methods is needed.

7.8 APPLICATIONS

7.8.1 Profile cutting of flat plate

The profile cutting of metals such as stainless steel, aluminium and copper alloys, which are difficult to machine by oxy-fuel gas techniques, is now a widespread industrial use for plasma machining, particularly when adapted for computer-numerical-control (CNC). Rates four times faster than oxy-fuel gas methods have been reported for plasma cutting of plates of thickness 6 to 25 mm. Commercial equipment utilizes about 200 kW of power.

7.8.2 Grooves

Table 7.6 summarizes the behaviour found with this application of plasma cutting: the dimensions of the grooves are clearly affected by arc power, traverse speed and angle and height of the plasma torch. Usually an orifice diameter of about 6.5 to 8.0 mm is used; cf. cutting for which a nozzle diameter not more than about 3 mm is used.

Grooves about 1.5 mm deep and 12.5 mm wide have been formed in stainless steel by PAM at metal removal rates up to 80 $mm^3 min^{-1}$ with equipment operated at 50 kW. These rates are about 10 to 30 times greater than those of conventional chipping and grinding. Plasma jet techniques can

Table 7.6 General effects of PAM process conditions on formation of grooves (After Lucey and Wylie, 1967)

Machining condition (increasing values)	*Groove condition*	
	Width	*Depth*
Arc power	Increase	Increase
Traverse speed	Decrease	Decrease
Torch angle	Decrease	Increase
Torch height	Increase	Decrease
Nozzle orifice	Increase	Increase

be used for similar groove formation in non-conducting materials, although the material removal rate is then reduced to about 30 $mm^3\,min^{-1}$.

Plasma machining can be used to produce grooves for subsequent welding. When butt-welds of high quality are required, a weld run can be made at one side of the joint, and then the reverse-side can be back-gouged by, for example, chipping, or grinding until sound weld metal is obtained.

7.8.3 Turning

Since the tool and workpiece do not come into contact in plasma machining, this method is particularly attractive for turning, especially with materials that are difficult to machine by conventional methods.

The plasma torch is held in a standard lathe in the same way as a conventional tool. As indicated in Fig. 7.9(a), the torch is usually mounted tangentially to the workpiece, at an angle of about 30°. Surface speeds are usually about 2 $m\,min^{-1}$ with a feed speed of about 5 $mm\,rev^{-1}$, yielding a surface finish of about 0.5 mm Ra (root to crest). As shown in Fig. 7.10, for turning of alloy steel, the metal removal rate rises with power used (the corresponding gas amounts consumed are also noted). The depth of cut can be controlled by either power or surface speed, as indicated earlier in Table 7.6.

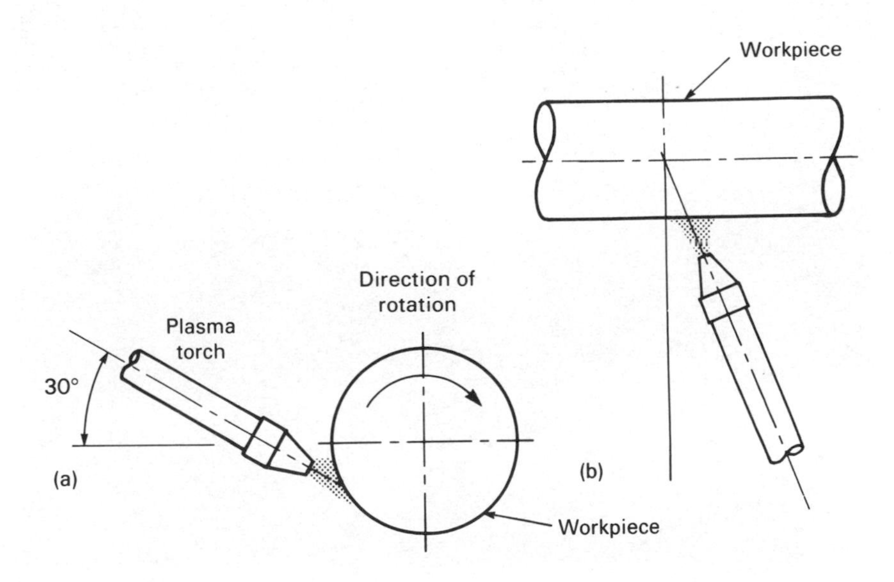

Fig. 7.9 Turning by plasma arc machining. (After Lucey and Wylie, 1967.)

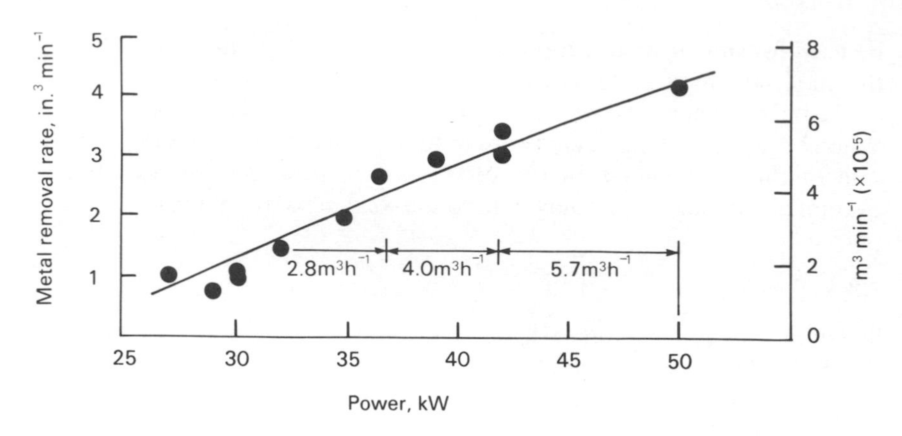

Fig. 7.10 Effect of power on metal removal rate. (Surface speed: 2.1 m min^{-1}.)

Fig. 7.11 A typical industrial plasma arc machine. (Courtesy of Trumpf Machine Tools Ltd.)

7.8.4 Underwater plasma machining

Recently work has been reported (Graham, 1980) on the reduction in noise, glare, and fumes associated with plasma machining, by immersing the plasma nozzle and workpiece in a depth of about 75 mm of water. Moreover nitrogen gas can then be readily employed as the plasma gas and expensive exhaust equipment is eliminated. Nitrogen is preferred to argon or argon/hydrogen mixtures for the plasma machining of mild steel plate, since it does not oxidize the surface produced and therefore the products of machining are soluble in the water. The slight consequential increase in acidity is reduced by appropriate replenishment by fresh water.

However, underwater plasma machining is hindered by the lower cutting speeds attained, and by operational difficulties with the needle-electrodes used. Nonetheless accuracies of 0.2 mm in 9 m have been claimed for numerically controlled underwater plasma machining. The use of the technique in seawater has been under investigation for some time, although little information is yet available.

The major drawbacks to the advancement of plasma machining lie with the large electrical power supplies needed for the process. Powers of 220 kW are needed to cut through 12 mm mild steel plate at 2.5 m min^{-1}. On the other hand the process readily lends itself to the computer-numerically controlled method. A direct numerical control (DNC) plasma machining installation for a bridge-building application has recently been publicized. Figure 7.11 shows a typical industrial plasma arc machining system.

BIBLIOGRAPHY

Anon. (1980) How Plasma Arc is Cutting into the Oxygen Market, *The Production Engineer*, **59** (11), November, 17–19.

Anon. (1983) Recent Developments in Plasma Technology, *Materials and Design*, **4**, 733–8.

Bellows, G. (1976) *Non-traditional Machining Guide 26 Newcomers for Production*, Metcut Research Associates Inc., Cincinnati, Ohio, pp. 54, 55.

Graham, A. (1980) Is Underwater Plasma Arc the Cutting Technique of the Future?, *The Production Engineer*, **59** (11), November.

Gustafsson, L. (1984) Research Shows that High-speed Plasma-arc Cutting of Gray Iron Engine Blocks is Possible, *Cutting Tool Eng.*, **36** (4), 24–6.

Lucey, J. A. and Wylie, F. S. (1967) Plasma Machining, *Proc. Conf. on Machinability*, The Iron and Steel Institute, London, 4–6 October, pp. 235–8.

Lunn, D. J. (1981) Nibbling Laser Cutting and Plasma Arc Cutting, *Tooling and Production*, **46** (12), 66–9.

Moss, A. R. and Sheward, J. A. (1970) The Arc Plasma Cutting of Non-metallic Materials, in *Electrical Methods of Machining Forming and Coating*, IEE Conf. Publ. no. 61, pp. 41–7.

8 Ultrasonic machining (USM)

8.1 INTRODUCTION

The prospects of using high frequency (about 70 kHz) sound waves for machining were noted as early as 1927 by Wood and Loomis; the first patents on this subject appeared in 1945, filed by Balamuth. Whilst investigating the ultrasonic grinding of abrasive powders Balamuth found that the surface of a container holding the abrasive suspension disintegrated when the tip of an ultrasonically vibrating transducer was placed close to it. Moreover, the shape of the cavity so produced accurately reproduced that of the tip of the transducer. A wide range of brittle materials, including glass, ceramic and diamond, could be effectively treated in this way.

The benefits to industry of this discovery were quickly recognized, and in the early 1950s the production of ultrasonic machine-tools began.

8.2 PRINCIPLES OF USM

The main parts of an ultrasonic machine are shown in Fig. 8.1. Its working may be understood in terms of the following elements.

8.2.1 Magnetostriction

A tool is oscillated at frequencies between 20 and 40 kHz, obtained by exploitation of an effect known as 'longitudinal magnetostriction'. With this phenomenon, a magnetic field undergoing variation at ultrasonic frequencies causes corresponding changes in the length of a ferromagnetic object placed within its region of influence. A magnetostriction transducer, such as that

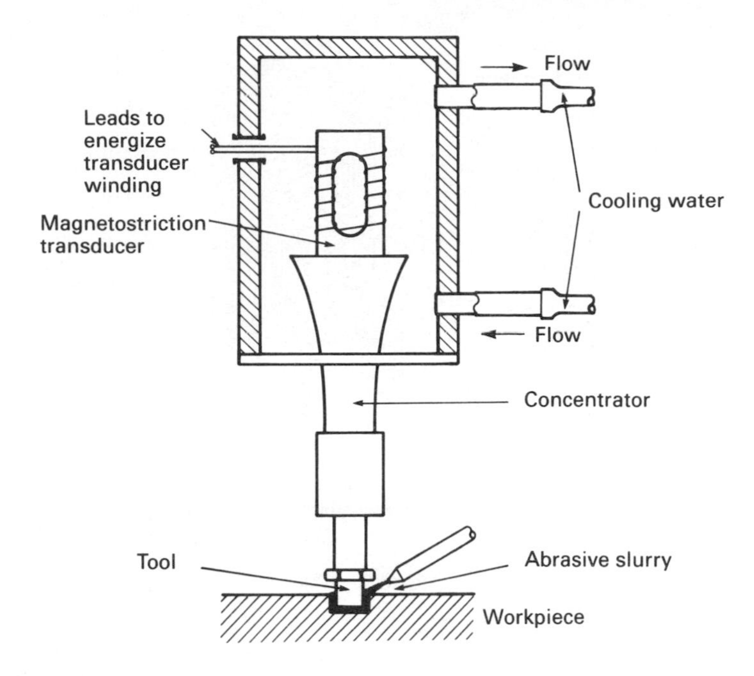

Fig. 8.1 Elements of ultrasonic machine. (After Kaczmarek, 1976.)

illustrated in Fig. 8.2, is used. The capacity of the magnetostrictor is described by its coefficient of magnetostrictive elongation

$$\epsilon_m = \frac{\Delta l}{l} \tag{8.1}$$

where Δl is the increment in length arising with the length l of the magnetostrictor core.

From thè typical relationship between this coefficient and the magnetic field intensity shown in Fig. 8.3 the elongation is noted to be independent of the sign of the magnetic field. Thus, a variation in magnetic field intensity causes changes in elongation at double that frequency. Although sinusoidal changes in elongation are not produced by this procedure, a means of procuring them can be adopted, as follows. The transducer is magnetized with direct current, in the polarizing winding; the changes in the magnetic field are then displaced to the rectilinear part of the characteristic curve of the dependence of magnetostriction capacity upon frequency, as indicated in Fig. 8.4. By this 'polarization' effect, sinusoidal changes in elongation are obtained.

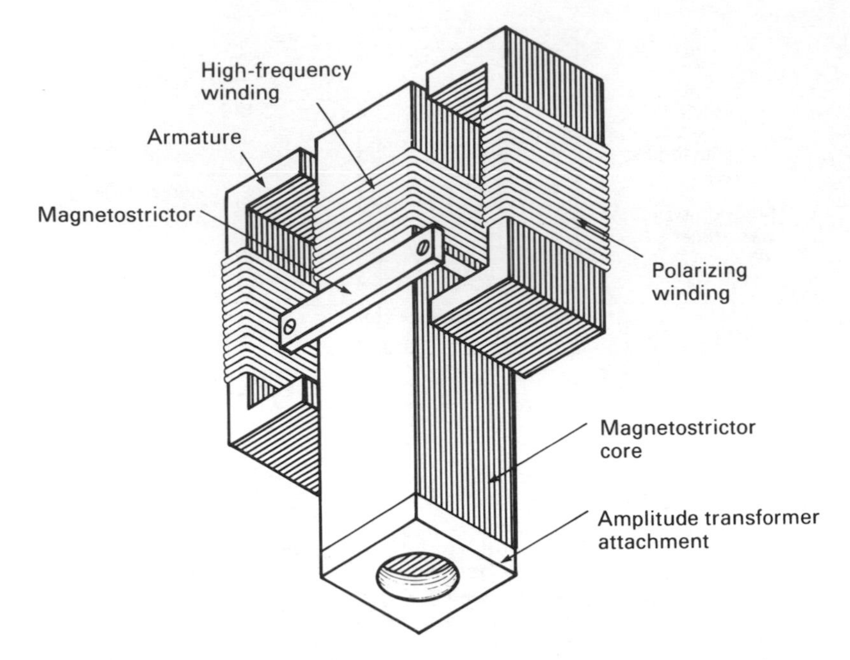

Fig. 8.2 Elements of magnetostriction transducer. (After Kaczmarek, 1976.)

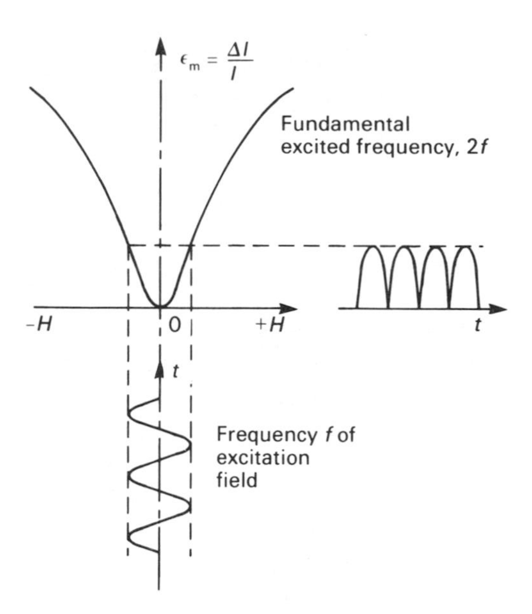

Fig. 8.3 Relationship between magnetostriction coefficient and magnetic field intensity, *H*; no magnetization. (After Kaczmarek, 1976.)

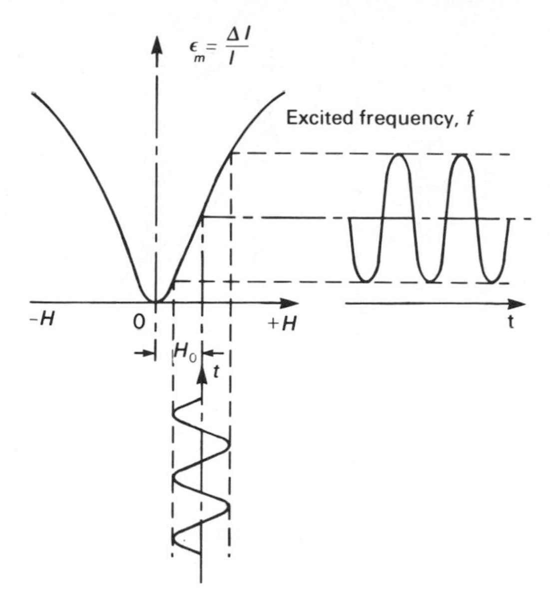

Fig. 8.4 Curves of magnetostriction excited by variable magnetic field; with magnetization. (After Kaczmarek, 1976.)

Figure 8.5 shows how the wave of elongation of a magnetostrictor varies along its length. If the mode of the elongation wave occurs at the centre of the length of the magnetostrictor, the maximum value of the amplitude of elongation will occur at a distance of a quarter of a wavelength from this centre. Therefore, the maximum elongation will correspond to a magnetostrictor length, l equal to half of the wavelength, λ (i.e. $l = 0.5\lambda$).

The wavelength λ is calculated from

$$\lambda = \frac{c}{f} = \frac{1}{f}\sqrt{\frac{E}{\rho}} \tag{8.2}$$

where c is the speed of sound (ms^{-1}) in the magnetostrictor material, f is the frequency (1/s) of the changes in the magnetic field, and E and ρ are respectively the Young's modulus (MPa) and density (kgm^{-3}) of the magnetostrictor material.

For condition (8.2) to be fulfilled, the resonance frequency f_r has to be chosen. It is given by

$$f_r = \frac{1}{2l}\sqrt{\frac{E}{\rho}} \tag{8.3}$$

where, as noted above, l is the length of the magnetostrictor.

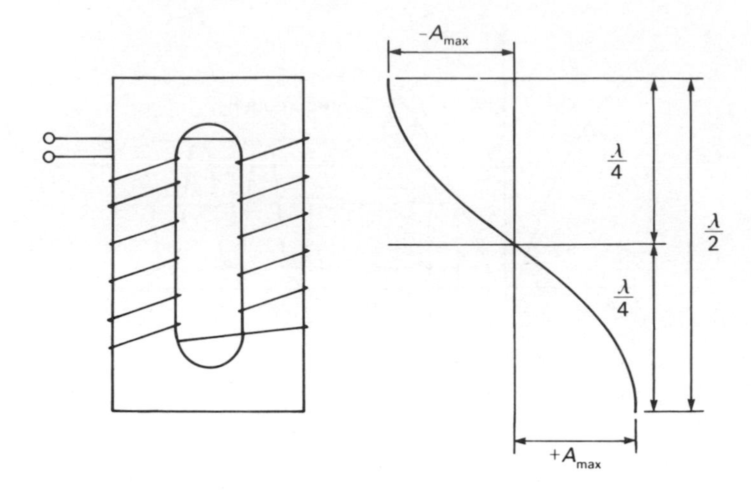

Fig. 8.5 Variation in wave of elongation along length of magnetostrictor. (After Kaczmarek, 1976.)

8.2.2 Amplitudes in USM

The elongations obtained from a magnetostrictor of length $l = 0.5\lambda$ operating in resonance are usually 0.001 to 0.1 μm. These elongations are too little to be of practical use.

The larger amplitudes, typically 0.01 mm, which are found to be necessary for efficient machining are obtained by fitting a waveguide focusing device, a 'concentrator' onto the output end of the tool. The concentrator is usually a cylindrically shaped metal rod. Thus the vibration-energy and amplitude are thereby correspondingly increased.

In order to obtain the maximum amplitude of vibration, the length of the concentrator is made a multiple of one-half of the wavelength of sound in the concentrator material. A resonance condition is then obtained. That is, the maximum amplitude of vibration is imparted to the tool.

8.2.3 Properties of magnetostrictive materials

Efficient operation of a USM system depends greatly on selection of magnetostrictive materials which have high values of the coefficient of magnetostrictive elongation in the saturated state, ϵ_{ms}. That is

$$\epsilon_{ms} = \frac{\Delta l_s}{l}$$

Also, since a conversion of magnetic to mechanical energy has to occur in USM, a satisfactory magnetostrictive material should possess high values of coefficient of magneto-mechanical coupling, k_r, given by

$$k_r = \sqrt{\frac{E_w}{E_m}}$$

where E_w and E_m are the magnitudes of, respectively, the mechanical and magnetic energy.

Table 8.1 indicates materials which are found to exhibit satisfactory properties.

Table 8.1 Properties of magnetostrictive materials (Data from Kaczmarek, 1976)

Material	*Coefficient of magnetostrictive elongation* ($\times 10^6$) ϵ_{ms}	*Coefficient of magnetomechanical coupling* k_r
Alfer (13% Al, 87% Fe)	40	0.28
Hypernik (50% Ni, 50% Fe)	25	0.20
Permalloy (40% Ni, 60% Fe)	25	0.17
Permendur (49% Co, 2% V, 49% Fe)	9	0.20

High fatigue strength for the concentrator material is also needed. Two popular relatively inexpensive choices are aluminium bronze and marine bronze with a fatigue strength of, respectively, 185 and 150 MNm^{-2}.

A drawback of magnetostrictive transducers is the high losses encountered with them. In many cases, the efficiency was as little as 55%. As a result the transducer heats up, and then has to be cooled. As an alternative, piezo-electric transformers were introduced into modern ultrasonic machines in the late 1960s, efficiencies of 90 to 95% being claimed. Tyrrell (1970) has described such a system. Briefly, an effect whereby a body can change its dimensions in response to an electric field is employed. Synthetic ceramics such as lead titanate-zirconate have been found to be most useful. These materials are composed of small particles bound together by sintering; this undergoes polarization by heating above the Curie point and placing it in an electric field such that orientation is preserved on cooling. A disc of the piezo-electric material which has a very high electromechanical conversion rating is sandwiched between two thick metal plates to form the ultrasonic horn. When a current of fixed frequency is fed to the horn the whole system is found to vibrate at some resonant frequency along the longitudinal axis; acoustically the motion is equal to one half a wavelength.

8.2.4 Materials for tools

The materials used for tool tips should have a high wear resistance and fatigue strength. The amount of wear is influenced by the workpiece material to be machined. Thus, as will be seen below, tool materials for USM vary between tungsten carbide, copper and silver steel, when say glass is the workpiece to silver and chromium-nickel steel for the machining of sintered carbides.

8.2.5 Static pressure of tool on workpiece

As indicated in Fig. 8.1, the tool is held against the workpiece by means of the static pressure exerted by a feed mechanism. This mechanism is fitted below the workpiece or above the tool. It is used to apply and sustain a force called the 'static load' between the tool and workpiece during machining, which has to overcome the resistance to the cutting action at the interface of the tool and the workpiece. For efficient motion, the feed mechanisms should have highly precise slideways, with low friction.

Of the many feed mechanisms available, the best utilize pneumatic systems, or periodic switching of a stepping motor or solenoid. Alternatively, compact spring-loaded systems sensitive to changes in cutting conditions have been used, admittedly with less efficiency. The simplest feed mechanism for USM employs a counterweight. However it is insensitive to changes in cutting.

8.2.6 Abrasive suspension

An abrasive suspension, usually composed of fine grit (size number 100 to 800), of boron nitride, aluminium oxide or silicon carbide in water, is circulated between the vibrating tool and workpiece. Under the combined effects of the constant static force on the tool or workpiece and the ultrasonic oscillations, the abrasive particles in the slurry are hammered into the workpiece, from which minute particles of material are removed by chipping.

The abrasive material should be at least as hard as the workpiece. Nonetheless the abrasive grit itself does become eroded during USM, so that a fresh supply has to be continually fed into the machining zone; this procedure also keeps the suspension cool during machining, and enables the debris to be efficiently removed.

8.3 MECHANISM OF MATERIAL REMOVAL

The way in which the abrasive particles remove material from the workpiece are still not completely understood, despite numerous investigations. However an explanation proposed by Kazantsev (1973) is a useful starting-point.

He first draws attention to the non-uniformity in size of the particles in the suspension. Because of this non-uniformity, as the tool is moved towards the workpiece surface it makes contact, partially, with some of the largest grains. These particles are hammered into the surfaces of both tool and workpiece. Kaczmarek suggests that the embedment of the abrasive grit in the tool surface gives rise only to plastic deformation there. However, at the workpiece actual disintegration occurs, by the chipping-out of a pocket at the surface.

Kremer, Saleh, Ghabrial and Moisan (1981) have examined the creation of chip pockets on the workpiece surface. They consider that these pockets are caused mainly by disintegration of the workpiece material, especially when it is brittle. The disintegration arises from the direct impact of the tool against the abrasive particles adjoining the workpiece surface.

These researchers, and Kazantsev, also report that cavitation-erosion also contributes to disintegration. The effect is especially marked when graphite is the workpiece material. In particular. Kazantsev points out that the random movement of the abrasive particles in the machining zone is mainly influenced by cavitation and the presence of air bubbles. The collapse of the cavitation bubbles in the abrasive suspension results in very high local pressures. Under the action of the associated shock waves on the abrasive particles, micro-cracks take place at the interface of the workpiece. The effects of successive rarefaction of the shock waves leads to chipping of particles from the workpiece.

The relative contributions made to material removal by the two effects, direct impact and cavitation erosion, have been found to vary with the operating conditions. Some workers claim that the latter effect accounts for less than 5% of the total volumetric removal rate.

Irrespective of the nature of these mechanisms, a brittle material is generally found to be more readily machined than a ductile one. For example the rate of machining of glass is much greater than that of a metal of comparable hardness. This disparity in USM behaviour is due to the difference in the plastic properties of the two materials. This influence of the plasticity of a material has led to the establishment of a 'brittleness criterion' for assessing the machineability of materials in USM.

8.4 BRITTLENESS CRITERION

The brittleness criterion τ_x is given by the ratio of shearing to breaking strength of a material. Efficiency of USM can then be assessed in terms of the brittleness criterion as shown in Table 8.2.

Table 8.2 Influence of brittleness criterion on efficiency of USM

Brittleness criterion τ_x	*Efficiency of USM*	*Typical materials*
$\tau_x > 2$	High	Glass, quartz, ceramic, diamond
$1 < \tau_x < 2$	Medium	Tempered steels, hard alloys
$0 < \tau_x < 1$	Low	Copper, lead, most steels

Owing to its low brittleness criterion, as shown in Table 8.2, steel is often adopted as the tool material in USM.

8.5 EFFECTS OF PROCESS CONDITIONS ON RATE OF USM

Although generally the more brittle a material, the greater is its machineability, no firm way exists of determining the rate of removal in terms of brittleness. A wide range of conditions, including process variables, properties of the tool and workpiece and of the abrasive slurry all play a part. A discussion of the main effects is therefore useful.

8.5.1 Amplitude of tool vibration

Firstly, the use of the resonant frequency of the acoustic system to give maximum amplitude at the tool tip has already been discussed above and has been noted by Kennedy and Grieve (1975). With a rise in amplitude of tool vibration the rate of USM generally increases, as confirmed by the experimental results given in Figs 8.6 and 8.7. The amplitude of tool vibration is generally recognized as having the greatest effect of all process variables on the rate of USM. However some workers including Markov (1977) suggest that vibration should be considered together with amplitude; they claim that both variables determine the velocity of the grit at the interface between the tool and workpiece. As expected, therefore, the machining rate rises along with velocity of the particle grit. Other properties of the grit also greatly influence the machining rate, as noted in the following section.

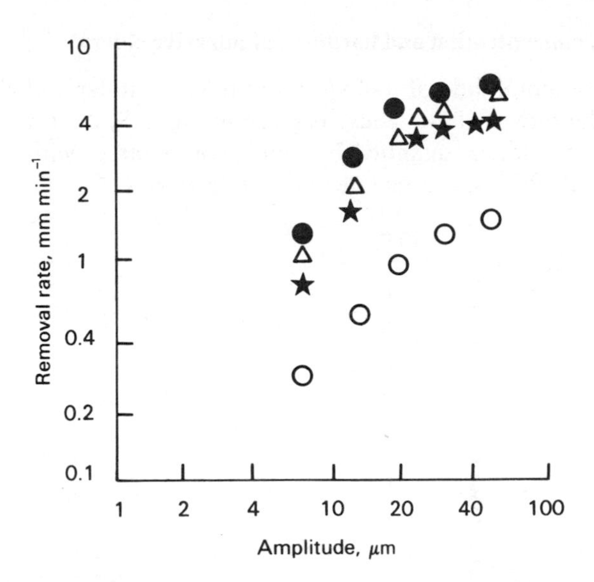

Fig. 8.6 Variation in removal rate with amplitude. (After Kremer *et al.*, 1981.) Glass workpiece, steel tool; abrasive 120 mesh B4C. Pressures: ● 0.20 MPa, △ 0.16 MPa, ★ 0.10 MPa, ○ 0.04 MPa.

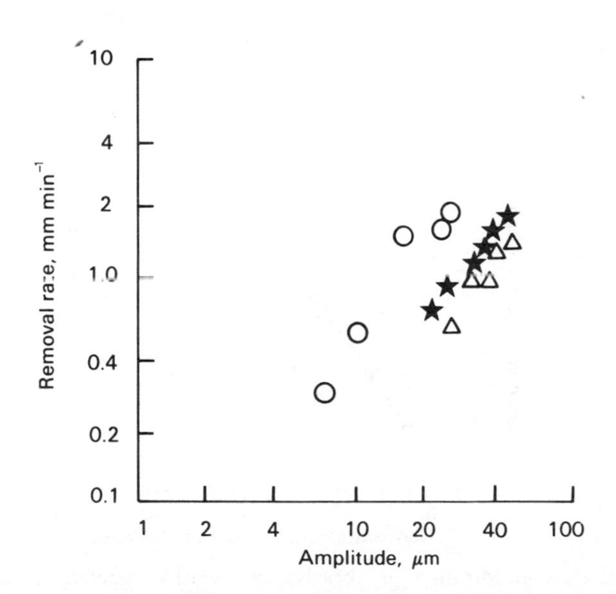

Fig. 8.7 Effect of amplitude of vibration on removal rate. (After Kremer *et al.*, 1981.) Hard steel workpiece, steel tool; abrasive 120 mesh B4C. Frequencies ○ 19 kHz, ★ 24.7 kHz, △ 43 kHz.

8.5.2 Size, concentration and hardness of abrasive slurry

Grit size and amplitude of tool vibration have a similar and closely related effect on the rate of USM. Thus, as given in Fig. 8.8, the rate rises with grit size until the latter quantity becomes comparable with the vibration amplitude, at which stage the rate of USM decreases.

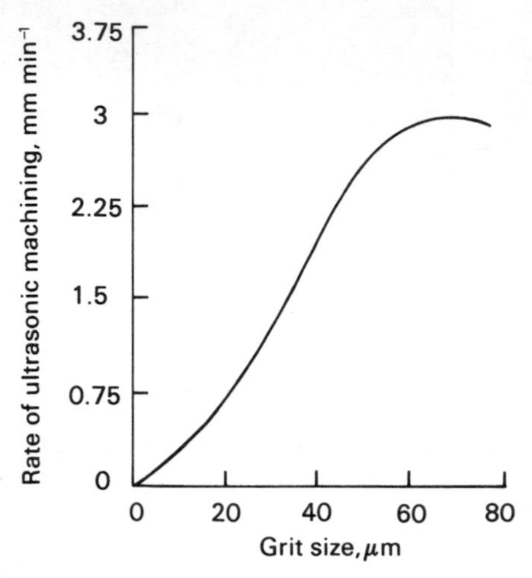

Fig. 8.8 Effect of grit size on rate of machining. (After Kazantsev, 1973.)

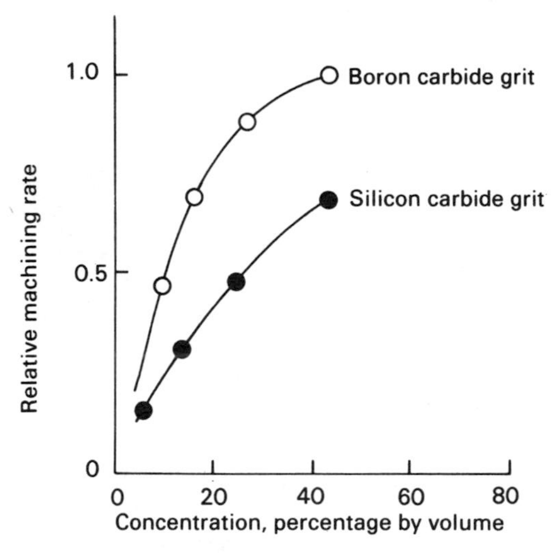

Fig. 8.9 Effect of concentration of abrasive on relative machining rate. (After Kennedy and Grieve, 1975.)

Note: maximum machining rate for boron carbide abrasive taken as unity; workpiece: soda glass.

As shown in Fig. 8.9 machining rates have been found to increase with the increase of concentration of the abrasive up to concentrations of 40% by volume.

The harder the abrasive the better the cutting action; in Table 8.3 confirmatory data are presented.

Table 8.3 Effects of hardness of abrasive on relative machining times. Grit size 60 μm (Data from Kazantsev, 1973)

Abrasive	*Microhardness (MPa)*	*Relative machining times*	
		Glass	*Alloy*
Boron carbide	42 200	1.0	40
Silicon carbide	31 400	1.2	120

The way in which the slurry is introduced to, and is distributed at, the machining zone is also found to affect machining rates. For example, when USM simply takes place in a bath of slurry, the effectiveness of the abrasive becomes progressively impeded and the machining rate decreases. In order to overcome this condition a jet of slurry can be directed at the tool/workpiece interface, as illustrated in Fig. 8.10. Alternatively the movement of the slurry over the machining zone can be controlled by suction, also noted in Fig. 8.10(b). The improved flow of the slurry results in enhanced rates of machining.

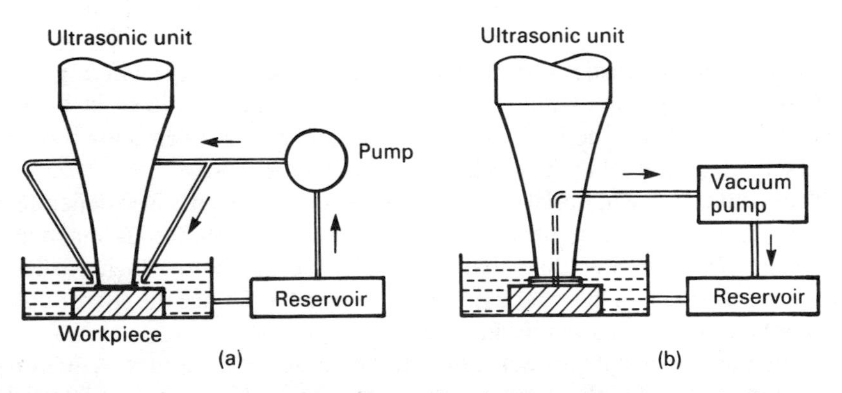

Fig. 8.10 Methods of injection of slurry. (After Kennedy and Grieve, 1975.)
(a) Jet flow.
(b) Suction.

8.5.3 Impact hardness of workpiece material

Conditions of machineability are also greatly influenced by the impact hardness of the workpiece material. The action whereby the abrasive grit becomes embedded in the surface of the material is similar to the indentation procedures used in the measurement of hardness.

In the assessment of this effect, useful information can be deduced from the analysis of the embedment of a single abrasive particle, which is assumed to have the shape of the segment of a sphere of diameter equal to the mean diameter of the grit. Experiments have shown that the impact indentation depth h is inversely proportional to the hardness H (Brinell number) of the material, according to the relationship:

$$H = \frac{F}{4hd} \tag{8.1}$$

where d is the diameter of the embedded sphere, and F is the force applied. Different types of workpiece material exhibit contrasting behaviour. Metallic and brittle materials are of particular interest.

The embedding of an abrasive particle in the surface of a metal workpiece is significantly different from that in a brittle material. In the case of metals, only plastic deformation and cold hardening occur, without disintegration. Almost immediate disintegration arises with brittle workpiece materials. A chip pocket of a pyramid-like shape with an angle of penetration of approximately 140° is formed in the workpiece. The angle remains virtually constant as the particle becomes increasingly embedded in the workpiece, although of course the size of the crater increases. Also, the average depth of indentation for brittle materials is found to be proportional to the force applied, unlike the behaviour for metals.

Another characteristic of brittle materials is the existence of a limiting force at which disintegration begins. This condition has been utilized in further studies in which the depth of the chipped pocket has been shown to be inversely proportional to the hardness of brittle materials.

The volume of the material removed in the process is clearly also dependent on the hardness. Since the hardness of most brittle materials machined ultrasonically is greater than that of metals, the depth to which particles become embedded in the machined surface is smaller. Hence the sharpness of the abrasive grit plays a significant role.

To that end, a 'disintegration parameter' is occasionally used; this quantity is defined as the ratio of the 'round-off' diameter of a sharply pointed particle to that (average) of the grains. For grain sizes varying from 30 to 300 μm, this parameter has been found to remain virtually steady with a value of 0.1.

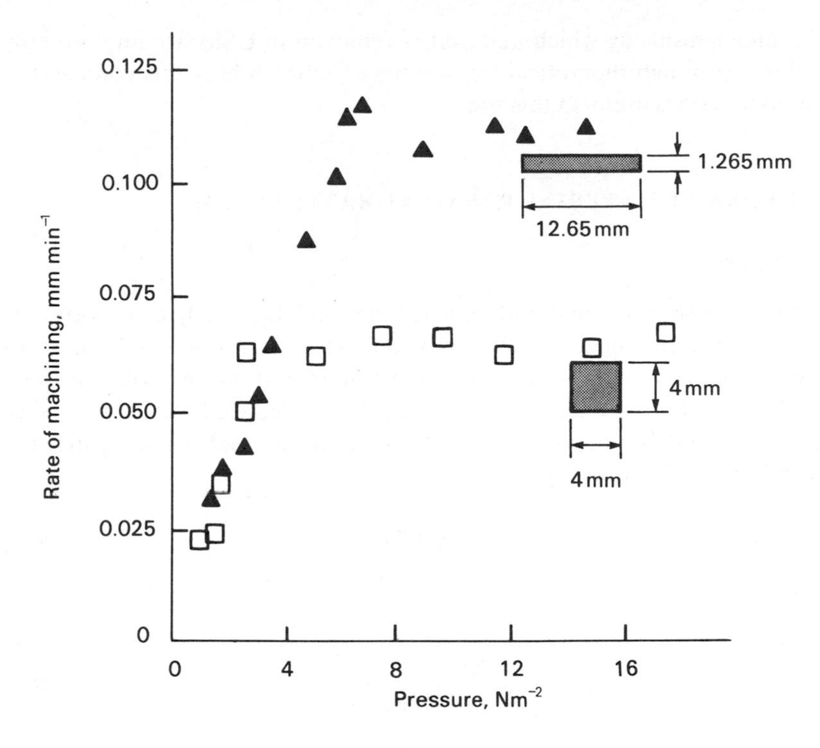

Fig. 8.11 Effect of tool shape on rate of machining (normalized values). (After Kennedy and Grieve, 1975.) Soda glass workpiece: tool areas 16 mm^2, □ square, ▲ rectangular section.

8.5.4 Effects of shape of tool on rate of machining

The rate of machining is also considered to be affected by the area and shape of the tool in USM. For example Kennedy and Grieve (1975) and Markov (1977) show that an increase in tool area reduces removal rate (mainly due to a consequential reduction in vibration amplitude). Kennedy and Grieve (1975) suggest that even if a constant amplitude of vibration could be achieved for different areas of tool, material removal rates would still decrease with rise in area, due to the problem of adequately distributing the abrasive slurry over the machining zone.

Figure 8.11 also confirms other general observations about the effects of tool shape on the machining rate in USM. For two tools of the same area, one of narrow rectangular shape yields a higher machining rate than another with square cross-section. Figure 8.11 also indicates that the machining rate rises with static pressure up to a limiting condition, beyond which no further increase occurs.

The mechanisms by which material is removed in USM are undoubtedly complex. Although theoretical treatments of USM reflect this complexity, some discussion is useful at this stage.

8.6 THEORY OF MATERIAL REMOVAL RATES IN USM

8.6.1 Analysis

A useful analysis of the material removal rates in USM has been reported by Kainth, Nandy and Singh (1979). They take account of the direct impact of abrasive grains upon the workpiece, basing their analysis on earlier work by Shaw (1956) and Rozenberg (1964). The latter deduced that the bulk of material in USM is removed by direct impact of the tool, showing that the material removal rate, $\dot{v}$, is given by

$$\dot{v} \propto [dh]^{3/2} Nf \tag{8.4}$$

where

$$h = \left[\frac{8F_s y_0 d}{\pi KHC(1+q)}\right]^{1/2} \tag{8.5}$$

Here d is the mean diameter of the grains, assumed to be spherical and identical; h is the depth of indentation, F_s is the static force, y_0 is the amplitude of vibration of the tool, H is the hardness of the workpiece, q is the ratio of hardness of the workpiece to that of the tool, C is the concentration of the abrasive slurry, and K is a constant of proportionality, evaluated from experiments. N is the total number of particles making impact per cycle, and f is the frequency of tool vibration.

Rozenberg (1964) derived a statistical distribution $F(d)$ for the abrasive grain size d based on experiments:

$$F(d) = 1.095\frac{N}{\bar{d}}\left[\left\{1-\left(\frac{d}{\bar{d}}-1\right)\right\}^2\right]^3 \tag{8.6}$$

where $\bar{d}$ is the mean diameter of the particles in the working gap, and N is their total number.

Kainth *et al.* investigated the effect of imhomogeneity in the size of spherical grains using the above statistical distribution given by Rozenberg on material removal rates. They first studied the mechanism of material removal in USM. As shown in Fig. 8.12, the tool descends from its mean position

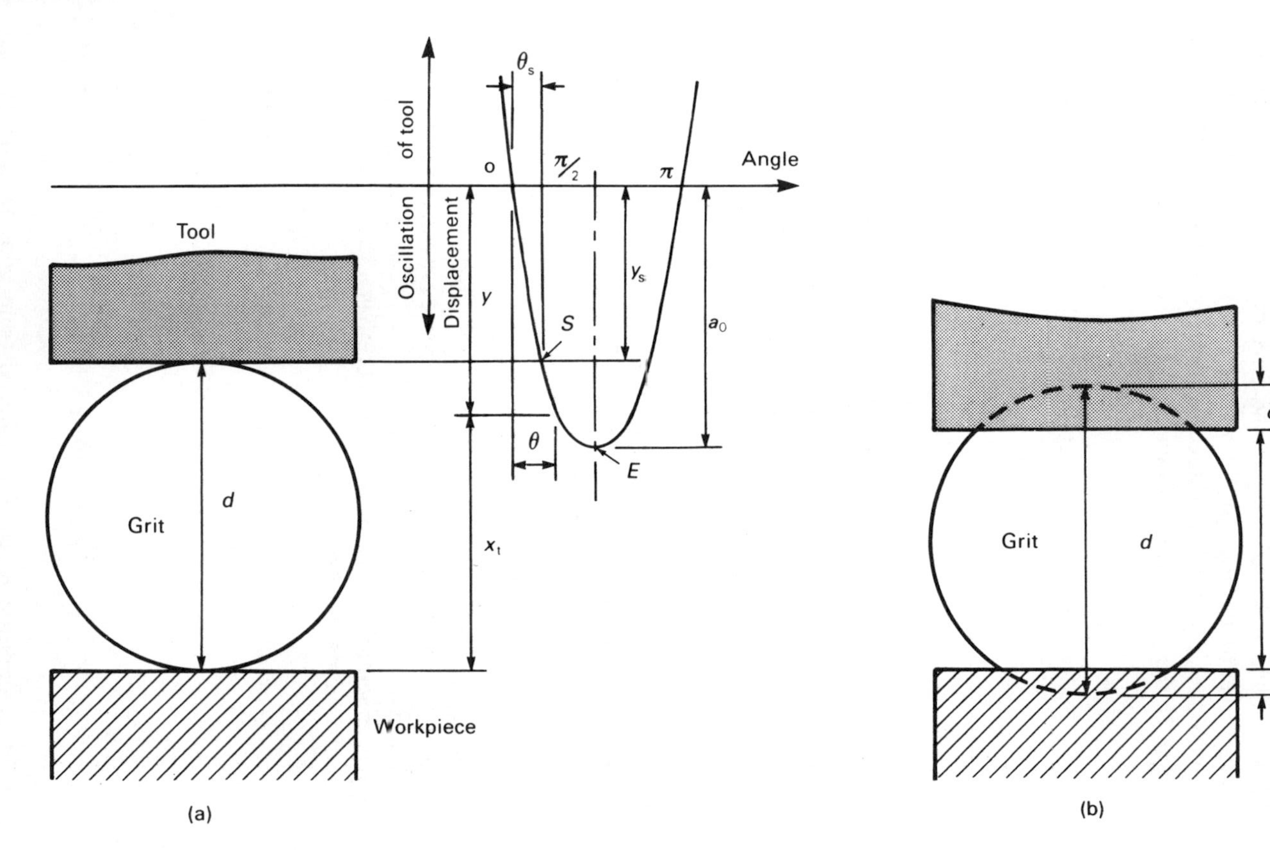

Fig. 8.12 Effect of tool and workpiece on impact with grain. (After Kainth *et al.*, 1979.)

touching the larger grains at position S. These are forced into the tool and workpiece as shown in Fig. 8.12(b). As the tool continues its motion downward, the grains undergo an increasing force on them, and may eventually fracture. Eventually, the tool terminates its downward movement at position E, a distance x from the workpiece surface, as shown in Fig. 8.12.

From Fig. 8.12 the total depth of penetration of an abrasive particle of diameter d is

$$\delta_t + \delta_w = (d - x) \tag{8.7}$$

δ_t and δ_w are respectively the indentation in the tool and workpiece, and x is the distance separating them.

The tool motion is sinusoidal with amplitude a_0. Thus

$$y = a_0 \sin\theta \tag{8.8}$$

where y is the vertical coordinate of distance and θ its angular position.

The motion of the tool is assumed to remain sinusoidal under the loaded condition, although it is noted that the force between the tool and the abrasive particles is effective only during a small portion of the cycle, and may alter the free sinusoidal motion.

The force of contact F_c acts only over the portion of the cycle between θ_s and $\pi/2$, as shown in Fig. 8.12 causing deformation of the tool and workpiece. Thus the total depth of indentation is given by

$$\delta_w + \delta_t = a_0 - y_s \tag{8.9}$$

where y_s is the distance moved downwards by the tool from its mean position, and the configuration of the workpiece and the abrasive grit, after indentation, is shown in Fig. 8.13.

The contact indentation zone in the workpiece has a radius r_w which for small values of δ_w/d may be expressed as

$$r_w = [\delta_w d]^{1/2} \tag{8.10}$$

From the shaded hemisphere region in Fig. 8.13, the volume V_0 fractured per grit is

$$V_0 = \frac{2}{3}\pi[\delta_w d]^{2/3} \tag{8.11}$$

The depth of indentation δ_w can be deduced from consideration of the

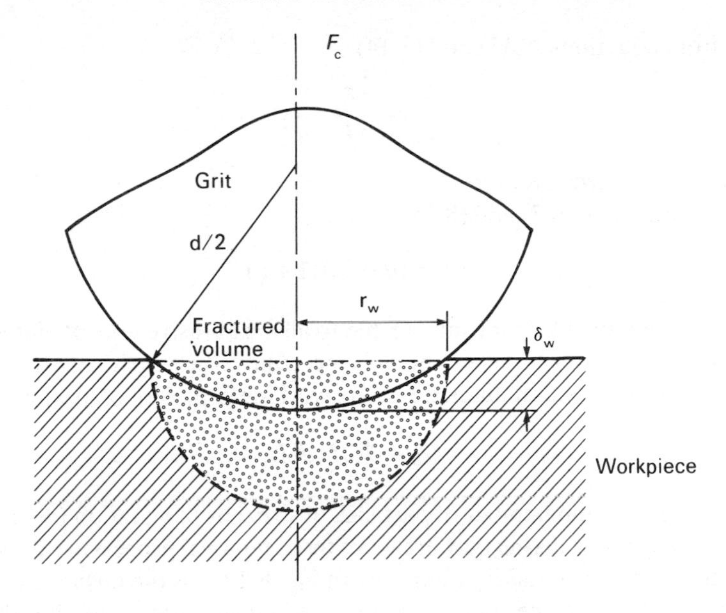

Fig. 8.13 Configuration of grit and workpiece. (After Kainth *et al.*, 1979.)

hardness H_t and H_w, respectively, of tool and workpiece. The brittle fracture hardness is defined by the average contact stress for fracture:

$$H_w = \frac{F_c}{\frac{\pi d}{2}[d - \{d^2 - (2r_w)^2\}^{1/2}]} \tag{8.12}$$

where F_c is the contact load on the grit.

For

$$\frac{\delta_w}{d} = \ll 1$$

and on substitution of the expressions for r_w from equation (8.10)

$$H_w = \frac{F_c}{\pi \delta_w d} \tag{8.13}$$

Similarly from consideration of indentation at the tool

$$F_c = \pi d H_w \delta_w \tag{8.14}$$

Then from equations (8.13) and (8.14)

$$\frac{\delta_t}{\delta_w} = \frac{H_w}{H_t} = q \tag{8.15}$$

where q is the hardness ratio.

From equations (8.7) and (8.15)

$$\delta_w = (d-x)/(1+q) \tag{8.16}$$

The rate of removal $V_d(\text{mm}^3\,\text{min}^{-1})$ due to all the abrasive grits of diameter d, may be expressed in terms of the volume removed per grit, the number of abrasive particles of that diameter in the working gap, and the frequency f:

$$V_d = V_0 F(d) f \qquad (8.16)$$

where $F(d)$ gives the number of particles of size d from equation (8.6).

The active grains taking part in the material removal process are those with diameters between x and d_m where, from Fig. 8.12, x is the distance between the tool and workpiece and d_m is the maximum diameter of the abrasive grains.

On consideration of all the effective grains, the total rate of material removal $\dot{v}$ is given by

$$\dot{v} = \int_x^{d_m} \frac{2}{3}\pi(\delta_w d)^{3/2} \left[1.095\frac{N}{\bar{d}}\left\{1-\left(\frac{d}{\bar{d}}-1\right)^2\right\}^3\right] f dd$$

On substitution for δ_w from equation (8.6) the total rate of removal becomes

$$\frac{2.29Nf}{(1+q)^{3/2}\bar{d}} \int_x^{d_m} [(d-x)d]\left[1-\left(\frac{d}{\bar{d}}-1\right)^2\right]^3 dd \tag{8.17}$$

In order to determine the removal rate from equation (8.17) the average number of particles N and the distance x have to be found. The same authors derive an expression for the former number, which is complicated and therefore not reproduced here. For the same reason their method of calculation of the distance x from the tool to the workpiece is not quoted here.

8.6.2 Theoretical effect of static load and amplitude of vibration on rate of machining

By solving their equations for N and x (total number of particles in the gap, and distance between tool and workpiece respectively), and substituting the values

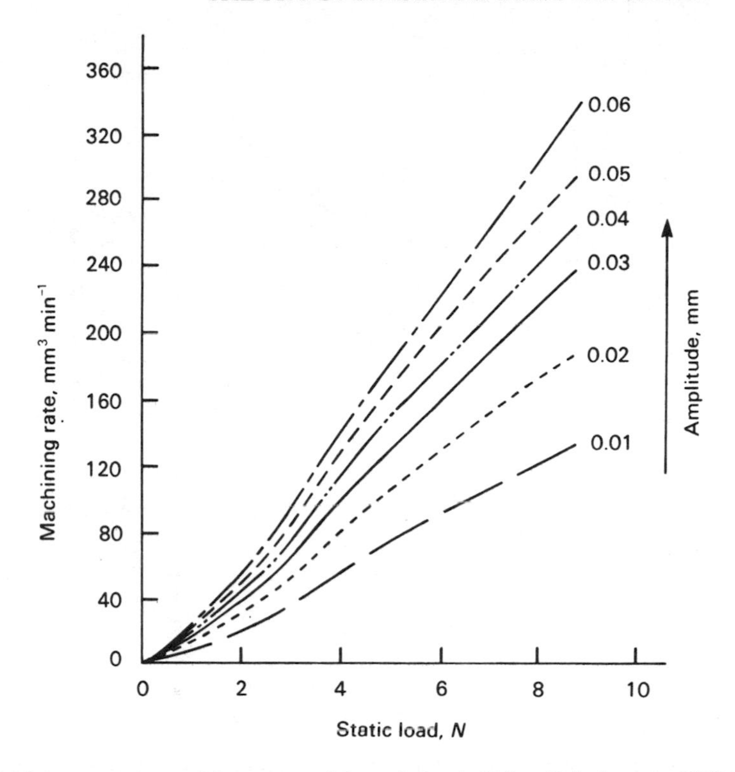

Fig. 8.14 Increase in machining rate with static load. (After Kainth *et al.*, 1979.) Glass workpiece; tool diameter: 12.7 mm; frequency: 25.5 kHz; concentration: 0.168; abrasive: boron carbide, 400 mesh.

in equation (8.17), Kainth *et al.* are able to find the dependence of machining rate on static load and amplitude. They then develop a computer program for the USM of glass of hardness 4600 MPa with a mild steel tool, of hardness 1470 MPa and of diameter 12.7 mm, vibrating at a frequency of 25.5 kHz, a 400 mesh boron carbide abrasive in water being used as the slurry. From the theoretical findings, Figs 8.14 and 8.15 show a rise in machining rate with static load. However Rozenberg (1964) has shown that in practice the machining rate rises initially and, after reaching an optimum value, it decreases with further increase in static load.

The theoretical increase in machining rate with amplitude shown in Fig. 8.15 is confirmed from the experiments of Nepiras, described by Rozenberg (1964), although the theoretical predictions are an order of magnitude higher than the experimental results (e.g. 32 mm^{-3} min^{-1} for soda glass).

Figure 8.16 shows that the analysis by Kainth *et al.* predicts that the machining rate rises linearly with abrasive grit size; in contrast Rozenberg

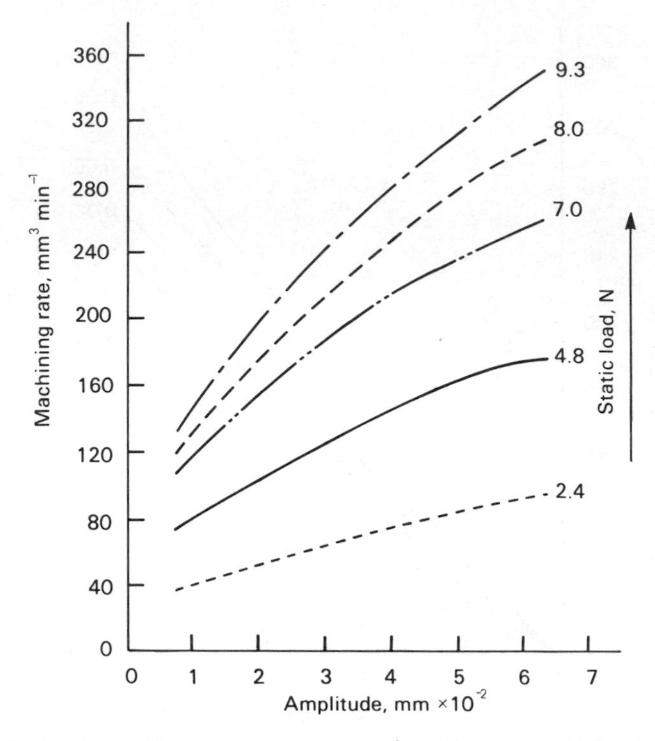

Fig. 8.15 Effect of amplitude and static load on machining rate. (After Kainth *et al.*, 1979.) Data as for Fig. 8.14.

(1964) reports that Fukomote found an optimum value of grain size which depends on the amplitude of tool oscillation.

Kainth and co-workers also discuss ways in which their analysis should be extended and modified to account more closely for experimentally determined behaviour.

8.7 SURFACE FINISH

Surface finish is closely associated with the rate of ultrasonic machining. Process conditions which influence the latter also have a significant effect on the finish obtained. Of the main process variables, the size of the abrasive grit is generally held to have most influence. As shown in Fig. 8.17, the larger the grit size, the finer becomes the surface finish. The amplitude of vibration has a smaller effect on surface finish. As the amplitude is raised, the surface roughness increases, since individual grains are pressed further into the

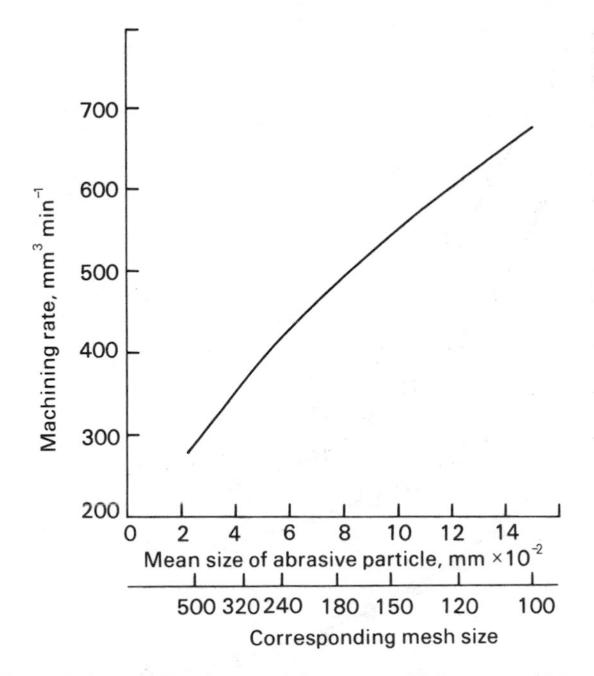

Fig. 8.16 Effect of size of abrasive particles on machining rate. (After Kainth *et al.*, 1979.) Data as for Fig. 8.14, except for amplitude: 0.0625 mm; static load: 9.1 N.

surface of the workpiece. Other process variables such as static load have little apparent effect on surface finish, even over a wide range of values.

Both tool and workpiece conditions have a bearing on the surface finish that can be obtained. Since the geometrical and surface features of the tool, including its surface irregularities, are reproduced on the workpiece, the finish on a tool should generally be finer than that required on the component. Widely varying finishes are obtained with different workpiece materials, as indicated also in Fig. 8.17. For example, Kremer *et al.* (1981) quote results in which the surface finish changed from 2.5 to 5.0 μm Ra for the USM of, respectively, graphite and glass under the same process conditions (with a 280 mesh boron carbide grit). Smoother surface finishes can be achieved when the viscosity of the liquid carrier for the abrasive is reduced (e.g. if machine oil is used instead of water).

Cavitation damage to the surface can occur, especially when deep cavities are ultrasonically machined. In these cases, complex changes in the process conditions occur in the machining zone within the body of the component; for example, the deeper the tool penetrates into the workpiece, the more difficult it becomes to replenish adequately the slurry in those deeper regions. The

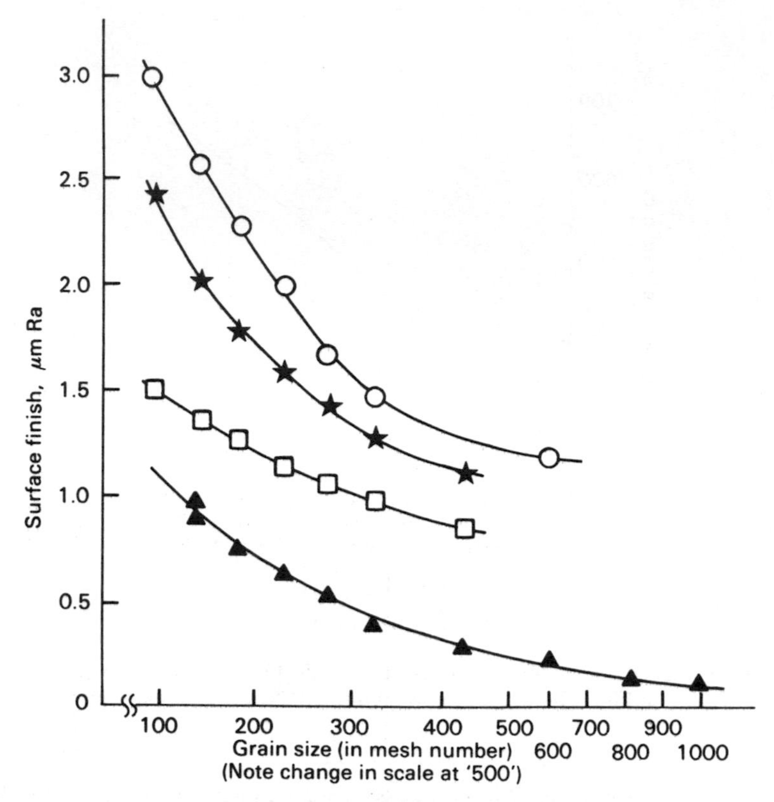

Fig. 8.17 Decrease in surface roughness with increase in grain size. (After Kennedy and Grieve, 1975.) Workpiece materials: ○ glass, ★ silicon semiconductor, □ ceramic, ▲ hard alloy steel.

average grit size therefore decreases and, as discussed above, the surface then becomes rougher.

Occasionally in USM, unavoidable changes occur in the amplitude of vibration, the static pressure and the area of cross-section of the zone of the workpiece being machined. Under such changes in the machining pattern, the surface finish can be deleteriously affected.

8.8 ACCURACY

8.8.1 Effects of process variables

The size of the abrasive grains has the principal effect on accuracy in USM. Accuracy however is also influenced by tool wear, transverse vibration and depth of cut. Analytic treatments of the dependence of accuracy on the main process variables is made very difficult by the complexity of the machining action. Consequently experimental observations are the main source from which general conclusions can be drawn.

8.8.2 Overcut

Many of these experimental data relate to hole-drilling by USM. For this application, 'overcut', that is, the clearance between the tool and the workpiece, is a frequently employed means of assessment of accuracy. According to Kremer *et al.* (1981), the overcut in USM is often about 1.5 times the mean grain size of the abrasive. From other results on the USM of glass and tungsten carbide, the size of overcut is considered to be about two to four times greater than the mean grain size. However, in general, the magnitude of the overcut depends on many conditions, including the type of workpiece material and the method of tool feed.

The overcut is usually greater at the region of entry of the tool to the workpiece than that at the exit. Conicity in the shape of the hole then arises, due to tool wear.

8.8.3 Conicity

Various procedures are used to reduce the conicity. For example, the use of higher static loads produces finer abrasives, which lessen the amount of tool wear and then, of course, conicity. The latter condition can be further reduced by direct injection of the slurry into the machining zone; it can sometimes then be eliminated entirely, by a further period of USM with fine abrasives. The conicity can also be lessened by use of tools which have negatively tapering walls.

8.8.4 Asymmetrical holes

The term 'out-of-roundness' is often used to describe asymmetrical holes. Asymmetry can arise due to lateral vibrations of the tool. These vibrations often occur when the centre-line of the tool is not perpendicular to the face of the workpiece, or when the acoustic parts of the machine are misaligned, or if the centre of gravity of the tool is off-centre of the vertical direction of motion.

From other work quoted by Kremer *et al.* (1981), the amount of 'out-of-roundness' is about 40 to 120 μm, and 20 to 60 μm, respectively, for glass and graphite. For the latter material, machining with a 20 mm diameter tool produced a conicity of approximately 0.2° for a depth of cut of 10 mm. The overcut obtained was about three times greater than the mean grain size of the boron carbide abrasive used, which had a mesh size of 280 to 600.

Little substantial information is available on the influence of tool wear on conicity in drilling, although Kennedy and Grieve (1975) claim that any effects on accuracy are small: they suggest that as the tool penetrates the workpiece, the undamaged portion of the tool corrects any errors produced in the earlier part of the machining operation.

8.8.5 Cavity sinking

In cavity sinking toolwear is much more significant. As USM proceeds the tool becomes increasingly tapered; this change in its shape is reproduced on the workpiece. The use of wear-resistant tool materials is therefore advocated. Kennedy and Grieve point out that the base of cavities cannot be machined flat, because the slurry becomes distributed unevenly across the machining face. Accuracies are then limited to about 0.05 mm.

8.9 APPLICATIONS

Although holes, slots and irregular shapes can be ultrasonically machined in any material, Kaczmarek (1976) suggests that the process should be applied mainly to brittle and hard materials involving surface areas of less than 1000 mm^2, where shallow cavities and cuts have to be produced.

8.9.1 Drilling

Typical results include the drilling of 0.15 mm diameter holes in 1 mm thick sapphire (De Barr, 1966). Bellows (1976) reports work in which a hole of diameter 6.25 mm and depth 125 mm was drilled in glass in 130 s. The tool wear and taper in the drilled hole were limited.

Drilling deep accurate holes in glass by USM has also been reported (Anon., 1973). Here a three-step process yielded accurate chip-free holes down to 1.59 mm diameter and up to 304 mm long by use of first a 12.7 mm long drill plated with 120 (USA notation) diamond grit, followed by similarly coated drills of length, respectively, 152 mm and 304 mm. The water-cooled drills were rotated at 4500 rev min^{-1}, and oscillated at 20 kHz, the amplitude at the tool-drill tip being about 0.0025 mm.

In another case, over 2000 holes 0.75 mm square in 1 mm thick carbon were drilled in less than ten minutes.

A wide range of other applications have been discussed by various workers. These include the threading of ceramics, lapping, broaching and deburring (Bellows, 1976; Graff, 1975). The latter author also discusses the use of USM in materials such as rock, bones and teeth. Kaczmarek (1976) describes USM for parting and cutting of semiconductor materials, such as germanium and quartz, commenting that the process yields high accuracy and machining rates in comparison with established methods with much smaller numbers of rejected components.

He also draws attention to applications for engraving on glass, and on hardened steel and sintered carbides (e.g. dies); production of spinning nozzles in ceramics, metals and minerals; and parting and machining of diamonds and other precious stones by use of diamond or borazon abrasive dust.

The use of USM for the manufacture of EDM electrodes has been claimed to reduce machining time from 20 hr to 30 min by copy-milling (Kremer *et al.*, 1983).

8.10 ULTRASONIC TWIST DRILLING

A modified version of USM that has attracted much attention is ultrasonic twist drilling, equipment for which is shown in Fig. 8.18. An ultrasonically activated drill bit is rotated against the workpiece, in a similar fashion to that of conventional twist drilling.

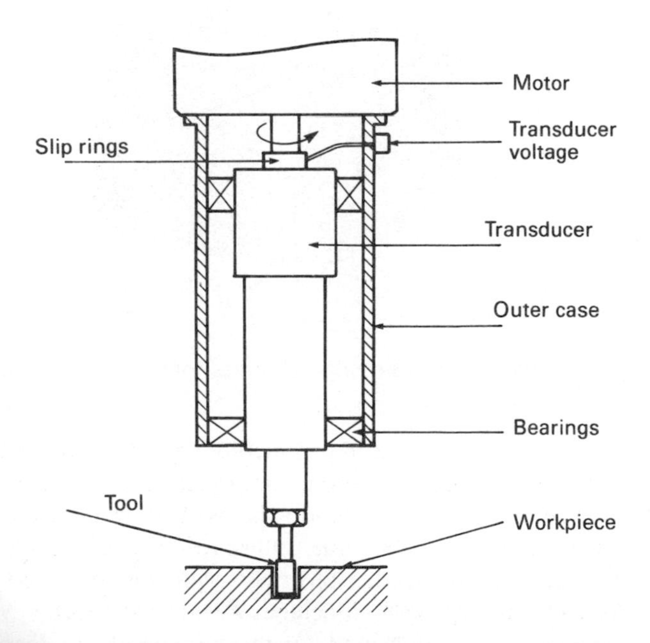

Fig. 8.18 Ultrasonic twist drill. (After Anon. 1973.)

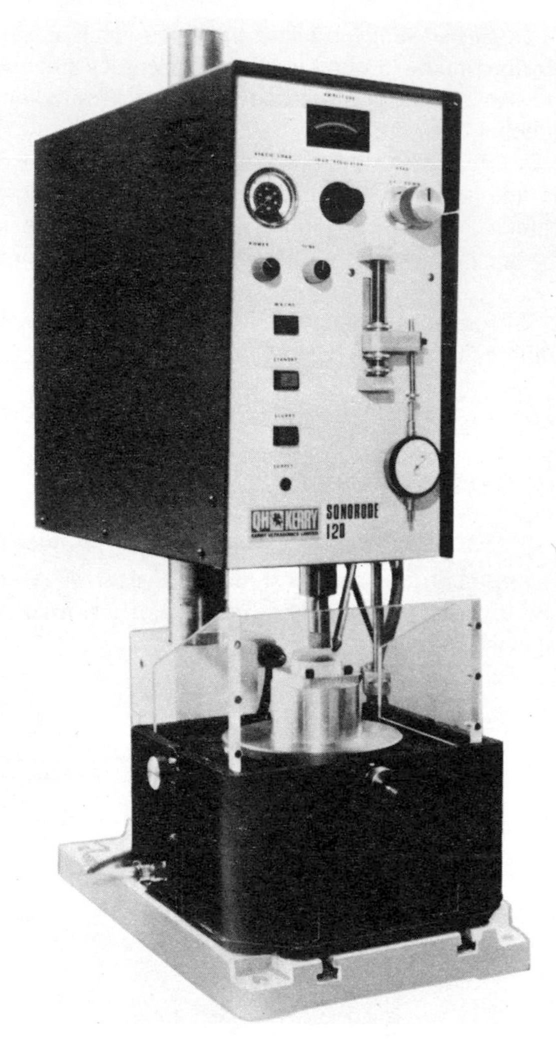

Fig. 8.19 An industrial ultrasonic machine. (Courtesy of Kerry Ultrasonics Ltd.)

An abrasive slurry is not used in this method. Most USM twist-drilling is carried out at 20 kHz and at vibration amplitudes of less than 1 mm. Holes 75 mm deep in ceramic, and 300 mm long holes, 1.5 mm in diameter in glass have been produced by this technique. Other results show that hole sizes as small as 80 μm can be produced. USM twist-drilling is restricted to circular holes, unlike conventional USM. Machines based on the former process are also more expensive and complicated. On the other hand, greater hole depths and two- to six-fold increase in cutting speed are attractive features of USM twist-drilling. The ability of the new technique to deal with intractable materials such as aluminium and titanium, as well as conventional ones, is also of value, since neither established USM nor conventional twist drilling can readily deal with either of these materials.

8.11 INDUSTRIAL ULTRASONIC MACHINE

Figure 8.19 shows the main features of an industrial ultrasonic machine.

BIBLIOGRAPHY

Anon. (1973) Drilling Deep Holes in Glass, *Ultrasonics*, May, p. 103.

Bellows, G. (1976) *Non-traditional Machining Guide, 26 Newcomers for Production*, Metcut Research Associates, Cincinnati, Ohio, p. 16.

Bellows, G. and Kohls, J. B. (1982) Drilling Without Drills, *American Machinist*, Special Report 743, p. 187.

Clouser, H. (1987) Ultrasonically Forming Intricate, Multiple, Graphite Electrodes, *EDM Digest*, **IX** (5), 12.

De Barr, A. E. (*c*. 1966) *Modern Metal Removal Techniques*.

Graff, K. F. (1975) Macrosonics in Industry, 5. Ultrasonic Machining, *Ultrasonics*, **13**, May, pp. 103–9.

Kaczmarek, J. (1976) *Principles of Machining by Cutting Abrasion and Erosion*, Peter Peregrinus Ltd, Stevenage, pp. 448–62.

Kainth, G. S., Nandy, A. and Singh, K. (1979) On the Mechanics of Material Removal in Ultrasonic Machining, *Int. J. Mach. Tool Des. and Res.*, **19**, 33–41.

Kazantsev, V. F. (1973) Ultrasonic Cutting, in *Physical Principles of Ultrasonic Technology*, Vol. 1 (Ed. L. D. Rozenberg), Plenum Press, New York, pp. 3–37.

Kennedy, D. C. and Grieve, R. J. (1975) Ultrasonic Machining – A Review, *The Production Engineer*, **54**, Sept., 481–6 and 103.

Kremer, D. *et al.* (1983) Ultrasonic Machining Improves EDM Technology, *Electromachining, Proceedings of the 7th Int. Symp.* (Ed. Prof. J. R. Crookall) 12–14 Apr. 83, Birmingham, UK, pp. 67–76.

Kremer, D., Saleh, S. M. *et al.* (1981) The State of the Art of Ultrasonic Machining, *Annals of the CIRP*, **30** (1), 107–10.

Markov, A. I., *et al.* (1977) Ultrasonic Drilling and Milling of Hard Non-metallic Materials with Diamond Tools, *Machines and Tooling*, **48** (9), 33–5.

Rozenberg, L. D. (1964) *Ultrasonic Cutting*, Consultants Bureau, New York. (Quoted in Kainth *et al.*, 1979.)

Shaw, M. C. (1956) *Microtechnic*, **10**, 165. (Quoted in Kainth *et al.*, 1979.)

Smith, T. F. (1973) Parameter Influence in Ultrasonic Machining, *Ultrasonics*, **11**, Sept., 196–8.

Soundararajan, V. and Radhakrishnan, V. (1986) An Experimental Investigation on the Basic Mechanism Involved in Ultrasonic Machining, *Int. Jnl Mach. Tool Des. Res.*, **26** (3), 307–32.

Tyrrell, W. R. (1970) Rotary Ultrasonic Machining, *Soc. Manufacturing Eng., USA*, Paper No. MR70-516, pp. 1–10.

9 Water-jet machining

9.1 INTRODUCTION

That materials can be cut with water has been known for many decades. However, the reliability and efficiency of the technique have reached satisfactory levels only in recent years. The key element in the process is a water jet, which travels at velocities as great as 900 ms^{-1} (approximately Mach 3). When the stream strikes a workpiece surface, material is rapidly removed by the erosive force of the water.

9.2 BASIC EQUIPMENT

These high velocities for the water jet are obtained from equipment illustrated in Fig. 9.1. A hydraulic pump powered from an electric motor, typically 30 kW, supplies oil at pressures as great as 117 bar in order to drive a reciprocating plunger pump, termed an intensifier (Fig. 9.2). This device plays a significant part in the process. It accepts water at low pressures, typically 4 bar, and expels it at far higher pressures, about 3800 bar, through an accumulator, which maintains the continuous flow of the high-pressure water, and also eliminates fluctuations, or spikes, in the pressure.

The accumulator relies on the compressibility of the water which is roughly 12% at 3800 bar in order to maintain a uniform discharge pressure and water jet velocity, when the intensifier piston changes in direction.

The water is then transported to the cutting head through high-pressure tubing, normally of diameter in the range 6 to 14 mm, the equipment being adapted for flexible movement of the cutting head. The cutting action is controlled, by means of a commercially available manually or remotely controlled valve specially designed for this purpose. See, for example, Norwood and Johnston (1984).

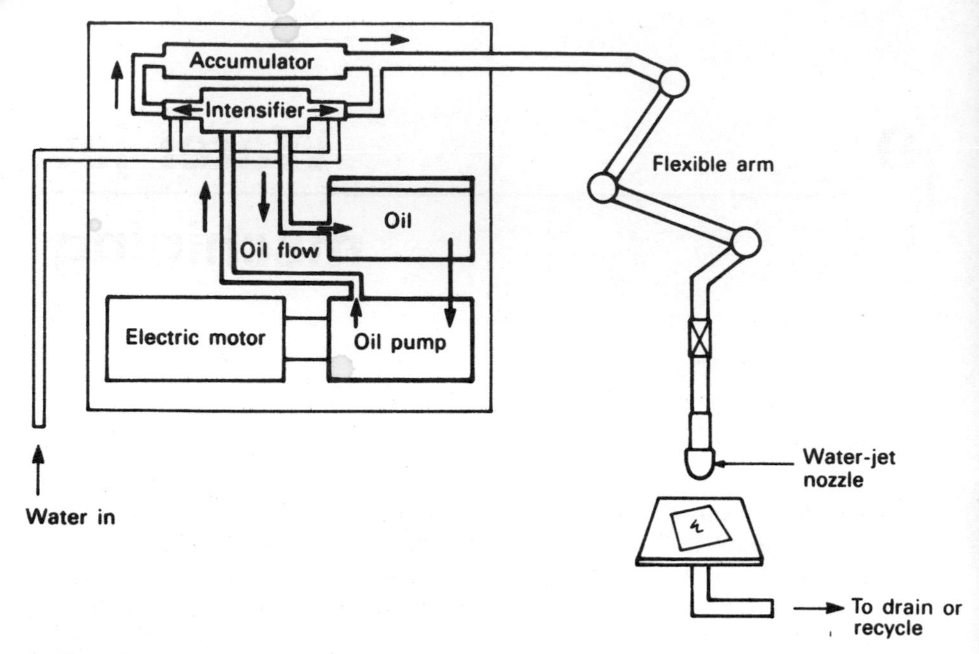

Fig. 9.1 Water jet machining system. (After Norwood and Johnston, 1984.)

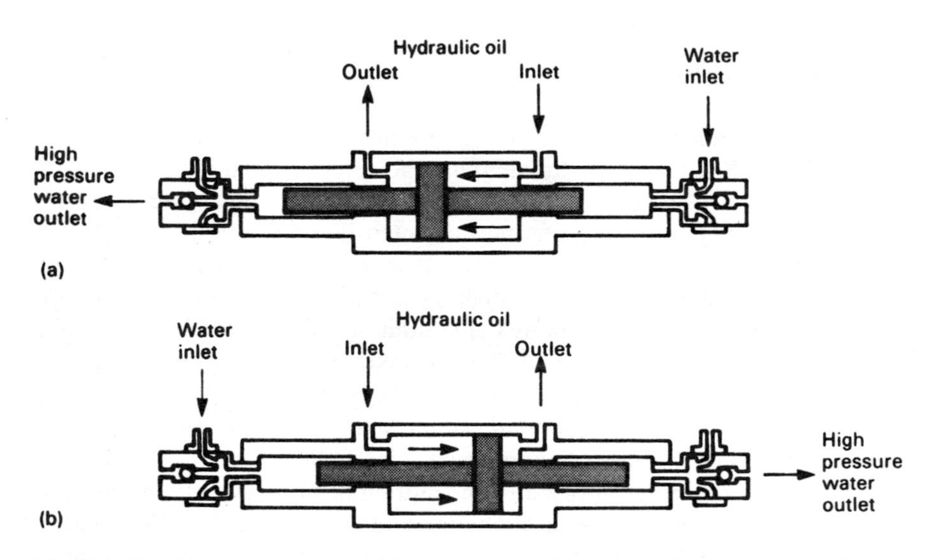

Fig. 9.2 Double-acting fluid intensifier. (After Norwood and Johnston, 1984.) Piston moving to
(a) left,
(b) right.

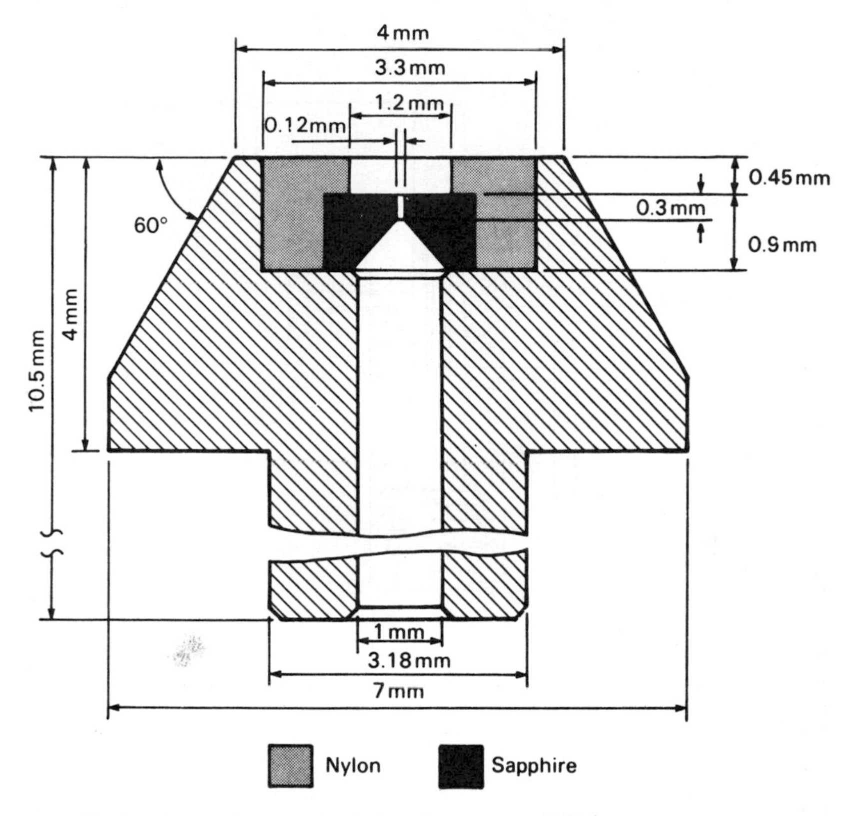

Fig. 9.3 Typical jet cutting nozzle. (After Engemann, 1981.)

The jet cutting nozzle is normally made from synthetic sapphire. An example is given in Fig. 9.3. About 200 hours of operation can be expected from a nozzle which becomes damaged by particles of dirt, and the accumulation of mineral deposits on the orifice due to excessive water hardness. Whilst a longer nozzle life can be obtained by simple multi-stage filtration which removes undissolved solids of size greater than 0.45 μm, more elaborate measures to extend nozzle life, such as de-ionization, are usually found to be too expensive to implement.

After the water jet has passed through the workpiece material, it is received in a 'catcher'. This device has two main functions. Firstly, if properly designed, it should reduce the noise levels associated with the reduction in the velocity of the water jet from super- (Mach 3) to sub-sonic levels. These noise levels can be as high as 105 dB. Secondly the catcher acts as a reservoir for collecting the machining debris entrained in the water jet.

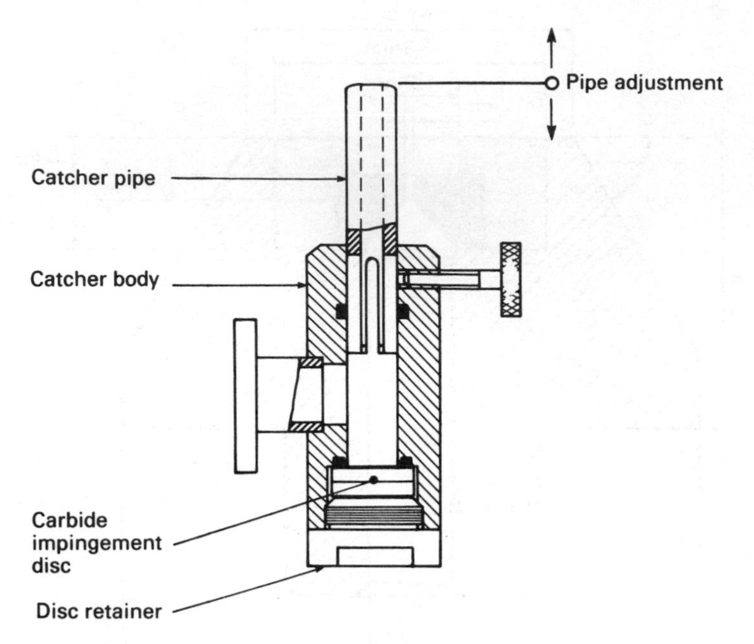

Fig. 9.4 Example of catcher. (After Norwood and Johnston, 1984.)

Table 9.1 Nozzle flow rates

Pump operating pressure bar	*Orifice diameter mm*	*Flow rate litre min^{-1}*
3450	0.13	0.64
	0.23	1.17
	0.30	2.65
	0.36	3.60
2760	0.15	0.57
	0.20	0.98
	0.30	2.27
	0.36	3.07
2070	0.15	0.49
	0.20	0.87
	0.30	1.97
	0.36	2.65

A simple catcher is shown in Fig. 9.4. It consists mainly of a small diameter pipe of variable length and a vacuum discharge port through which the slurry of cutting water and machining debris is disposed.

For some applications in which the operators are remote from the cutting zone, a catcher basin is used for collection of the water and disposal of the waste material.

The gap between the jet and the workpiece is often termed the 'stand-off' distance; a distance of 2.5 to 6 mm is usual, although for some solid materials, such as those used in printed circuit boards, it may be increased to 13 to 19 mm.

The quality of cutting is usually found to improve with increase in pressure, by widening the diameter of the jet, and by lowering the traversing speed. These procedures also enable materials of greater thickness and densities to be cut.

Table 9.1 presents typical values of pump operating pressures, orifice sizes and corresponding flow-rates.

9.3 THEORETICAL CONSIDERATIONS

Norwood and Johnston have proposed that Bernoulli's equation may be used to provide a reasonable estimate of the velocity v of the water cutting jet:

$$v = (2p/\rho)^{1/2} \tag{9.1}$$

where p is the pressure, and ρ is the average density of the cutting fluid. They also provide the following expression for evaluating the volumetric flow-rate Q:

$$Q = C_D(p/4)D^2(2p/\rho)^{1/2} \tag{9.2}$$

where C_D is the orifice coefficient, typically 0.7 for a new orifice of diameter D.

9.4 ADVANTAGES OF WJM

Norwood and Johnston (1984) list 16 advantages of WJM over alternative methods, including routers, saws, knives, dies and lasers. These main attractions include

1. no wear of the 'tool', which does not need sharpening;
2. no heat development;

3. no chip or dust formation or foreign matter left in the workpiece or component;
4. cut surfaces are clean;
5. little attention needed;
6. lightweight, low-cost tooling;
7. omni-directional cutting is possible.

On the other hand, high costs are still a major impediment to the advancement of this technique.

9.5 APPLICATIONS

9.5.1 Cutting

Early applications were limited to the cutting of fibreglass and corrugated board due to the comparatively primitive equipment available. The development of new items of equipment such as automatic jet sensors and quick-acting valves has made possible many new applications for the process.

9.5.2 Machining of fibre-reinforced plastics

Konig, Wulf, Graß and Willerscheid (1985) have shown that in the water-jet cutting of fibre-reinforced plastics, thermal material damage is negligible. The tool being effectively pointed provides for accurate cutting of contours. A main drawback is the deflection of the water jet if it hits an obstacle, such as a fibre embedded in the matrix; the jet then cuts neighbouring soft material and the fibre can protrude after machining. The feed rate attainable is often limited by the surface quality required. Table 9.2 shows typical limiting feed rates for water-jet cutting of fibre-reinforced plastics.

Table 9.2 Limiting feed rate for water-jet cutting. Data: Water pressure 3500 bar, nozzle diameter 0.18 mm, standoff distance 2 mm (from Konig *et al.*, 1985)

Material	*Thickness (mm)*	*Feed rate ($m\ min^{-1}$)*
GFRP	2.2	1.8 to 6.0
Laminate	3.0	1.4 to 5.0
	5.0	0.4 to 3.0
AFRP	1.0	10.0
(Weave)	2.0	2.4 to 4.0

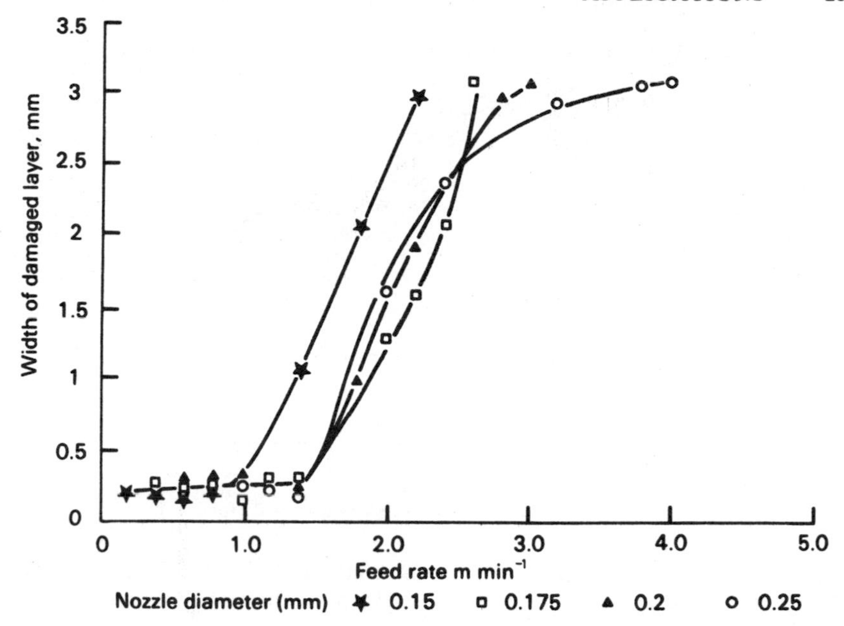

Fig. 9.5 Effect of feed rate on width of damaged layer. (After Konig *et al.*, 1985.) Material: fibre-reinforced plastic; thickness: 3 mm; pressure: 350 MPa; stand-off distance: 2 mm.

The pumping pressure, which should be kept above 300 MPa, is again considered to be a very significant factor. A rise in pressure increases the limiting feed rate, and improves the surface quality. Although the power of the water jet increases with the square of the nozzle diameter, the maximum feed rate rises only linearly with the latter quantity, as illustrated in Fig. 9.5.

The water jet loses velocity as it penetrates the material, causing a decrease in surface quality towards the exit. From the experiments performed by Konig *et al.* (1985) generally the cut surface consists of a smooth zone with only light vertical grooves on the side of the jet entrance together with a rougher exit containing larger grooves. The transition between the two zones is abrupt.

Engemann (1981) also reports on water-jet cutting of fibre-reinforced composite materials, showing that an increase in the distance between the jet and specimen surface reduces the depth that can be cut; see Fig. 9.6. He describes the water-jet trimming of a helicopter roof made from a polycarbonate material, drawing attention to robotic control of the water jet.

9.5.3 Water-jet cutting of rock

In mining and quarrying, slots more than 2 m in depth often have to be produced, for instance for cutting tunnel profiles prior to drilling and blasting,

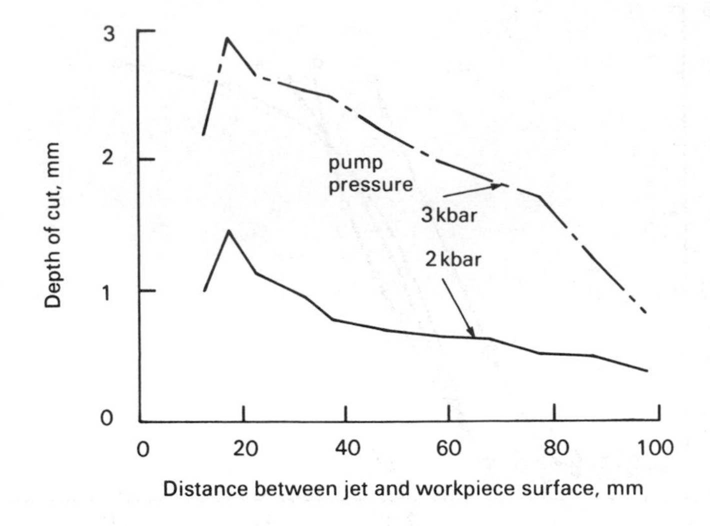

Fig. 9.6 Effect of jet/workpiece distance on depth of cut. (After Engemann, 1981.) Diameter of jet: 0.12 mm; feed-rate of jet head across workpiece: 1.1 mm s^{-1}.

and cutting trenches in rock and concrete. The single, small-diameter water jet methods are of limited use, since the force needed for cutting becomes dissipated on interaction with the wall. Also deep slots have to be sufficiently wide for the water-jet nozzle to enter. Nonetheless, if that can be achieved a satisfactory 'stand-off' distance between the nozzle and the material face can then normally be obtained for useful cutting-rates to be achieved. Wide slots can be obtained by use of a rotating nozzle with jets angled to yield the desired slot width. Alternatively a nozzle with angled jets can be oscillated, so that it covers the area of the slot. For applications like coring, a small nozzle with multiple angled jets is used, its motion being needed only in the direction of the slot.

Reichman and Cheung (1978) report the water-jet cutting of a slot, 51 mm deep in granite by two 0.3 mm oscillating jets at 275 MPa. The cut was made in 14 passes at a feed-rate of 25.4 mm s^{-1}.

An oscillating nozzle system operating at a feed rate of 25.4 mm s^{-1} and a pressure of 172 MPa, with the stand-off distance adjusted after every other pass, was used to cut a 178 mm deep slot in sandstone. Comparatively constant wall spacings were obtained (although the width is not given).

9.5.4 Water-jet deburring

The removal of surface irregularities or burrs by water-jet methods has become increasingly attractive. Three techniques are available. In 'wet-blasting' a low-pressure (up to 4.2 bar) pump is used to accelerate large volumes of water-abrasive mixtures up to velocities of 30 ms^{-1}. Sometimes up to 30% of abrasives are added, preferably silicon carbide, corundum or glass beads, of grain size 10 to 150 μm. In this case material is removed by the erosive effects of the abrasive with the water acting as its carrier and dampening the effects of its impact on the surface. Thiel, Przyklenk and Schlatter (1984) describe the water-jet deburring of irregularities 0.35 mm high and 0.02 mm wide left on a steel component after grinding. At a water pressure of 3.5 bar with an abrasive of glass beads of diameters between 105 and 210 μm, the component was effectively deburred in about 30 s.

Sometimes compressed air is introduced into the water jet, causing its atomization, in order to enhance the deburring action.

Water-jet deburring can also be achieved without the use of abrasives, the burrs being broken off by the impact of the water. Plunger pumps capable of pressures up to 1000 bar are used, the technique being particularly applicable for large burrs, and for those in narrow cavities. For example a 3 mm high burr formed in a 12 mm wide hole drilled in a hollow molybdenum-chromium steel shaft was removed in 15 s by a water jet operated at 700 bar and 27 litre min^{-1}, the nozzle 'stand-off' distance being 1.5 mm (Thiel *et al.*, 1984).

A similar system of water-jet deburring (with abrasives added) by injector nozzles is claimed to remove irregularities very rapidly from hardened steel. Two cross-holes, of 10 mm and of 5 mm diameter formed in an aluminium component were deburred in 1 s, with an abrasive of broken glass added to a water jet operating at 400 bar.

Very high-pressure, low flow-rate jet cutting with pump pressures of 4000 bar and flow rates of 2.5 litre min^{-1} is claimed to be well suited for the removal of burrs from plastic and rubber (non-metallic) components.

9.5.5 Water-jet cutting of printed circuit boards

Norwood and Johnston (1984) have discussed the cutting of printed circuit boards (PCB) by WJM, reporting speeds of more than 8 m min^{-1}, to accuracies of ±0.13 mm. They point out that the water jet has a very small diameter, and therefore can pass close to components mounted near to the edge of the board. No detrimental effect of the water from the process on the PCB is reported. These authors describe the cutting of boards of various shapes, for use in portable radios and cassette players, master boards being loaded onto an automatic machine, from which individual PCBs are cut by WJM, as shown in Fig. 9.7.

For this kind of application water-jet systems with as many as 14 *X–Y* cutting stations in one plant are now being used.

A typical water-jet machine is shown in Fig. 9.8. This industrial unit is

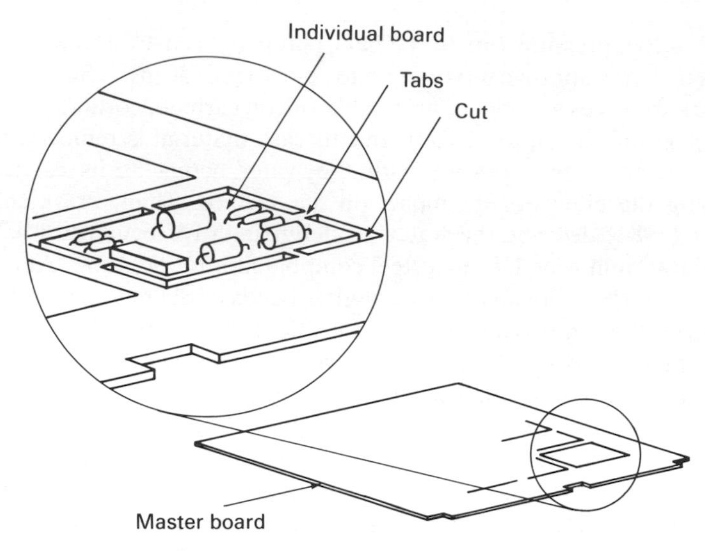

Fig. 9.7 Printed circuit board cut by WJM. (After Norwood and Johnston, 1984.)

Fig. 9.8 A typical water-jet machine. (Courtesy of Jetin Industrial Ltd.)

capable of profiling by manual or optical line following; the machines can be adapted for numerical or computer control.

BIBLIOGRAPHY

Atkey, M. (1983) Water Jet Cutting: a Production Tool, *Mach. Prod. Eng.*, **141** (3631), 18–19.

Beutin, E. F., Erdmann-Jesnitzer, F. and Louis, H. (1974) Material Behaviour in the Case of High-speed Liquid Jet Attacks, *Paper C1 of the 2nd Int. Symp. on Jet Cutting Technology*, Cambridge, pp. C1–18.

De Meis, R. (1986) Cutting Metal with Water, *Aerospace America*, **24** (3), 22–3.

Engemann, B. K. (1981) Water Jet Cutting of Fibre Reinforced Composite Materials, *Industrial and Production Engineering* (3) 162–4.

Haferkamp, H., Louis, H. and Schikorr, W. (1984) Precise Cutting of High Performance Thermoplastics, *Paper G2 from the 7th Int. Symp. on Jet Cutting Technology*, pp. 353–68.

Konig, W. and Wulf, C. H. (1984) The Influence of Cutting Parameters on Jet Forces and the Geometry of the Kerf, *Paper D2 from the 7th Int. Symp. on Jet Cutting Technology*, Ottawa, Canada, pp. 179–91.

Konig, W., Wulf, C. H., Graß, P. and Willerscheid, H. (1985) Machining of Fibre Reinforced Plastics, *Annals of the CIRP*, **34** (2), 537–48.

Labus, T. J. (1978) Cutting and Drilling of Composites Using High Pressure Water Jets, *Paper G2 of the 4th Int. Symp. on Jet Cutting Technology*, University of Kent at Canterbury, pp. G2, 9–18.

Norwood, J. A. and Johnston, C. E. (1984) New Adaptions and Applications for Waterknife Cutting, *Paper G3 from the 7th Int. Symp. on Jet Cutting Technology*, pp. 369–88.

Reichman, J. M. and Cheung, J. B. (1978) Waterjet Cutting of Deep-kerfs, *Paper E2 of the 4th Int. Symp. on Jet Cutting Technology*, University of Kent at Canterbury, pp. E2, 11–28.

Thiel, R., Przyklenk, K. and Schlatter, M. (1984) Deburring with Water, *Paper G1 from the 7th Int. Symp on Jet Cutting Technology*, pp. 337–52.

Verma, A. P. (1985) Basic Mechanics of Abrasive Jet Machinery, *J. Inst. Eng. India*, Part PE2, **66**, 74–81.

Yanaida, K. (1974) Flow Characteristics of Water Jets, *Paper A2 of the 2nd Int. Symp. on Jet Cutting Technology*, Cambridge, pp. A2, 19–32.

10 Specialized methods of machining

10.1 INTRODUCTION

In this chapter a range of advanced methods of material removal, some of which have only very recently been devised, are discussed. First to be described however, 'abrasive jet machining', has some similarity to the water-jet method discussed in Chapter 9. Then chemical machining is investigated. This well-established technique has been a pioneer for some of the other newer methods studied in subsequent sections.

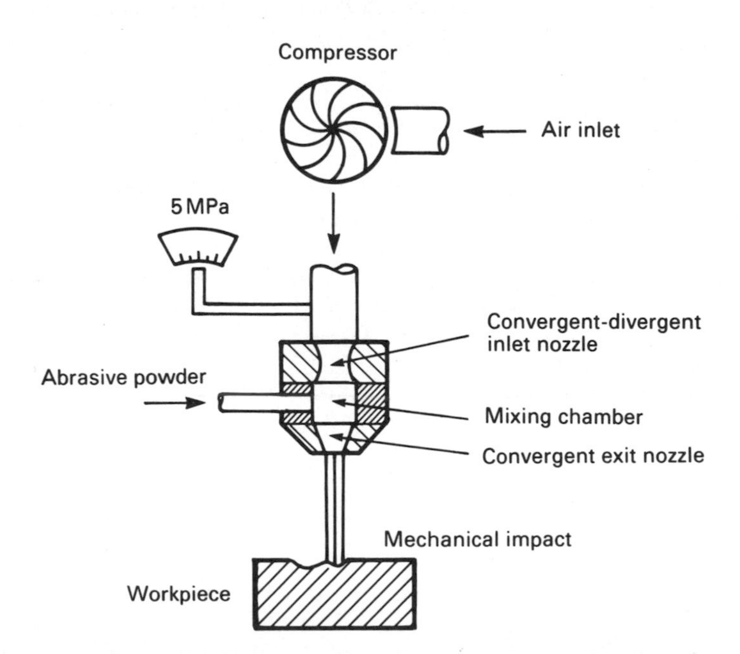

Fig. 10.1 Abrasive jet machining process. (After Snoeys *et al.*, 1986.)

10.2 ABRASIVE JET MACHINING (AJM)

Abrasive particles, usually aluminium oxide or silicon carbide powders about 60 μm in diameter, are carried in a high-velocity air jet at a pressure typically 5 MPa and are used to erode materials. The air jet is passed through an orifice of diameter 0.15 to 2.0 mm in a tungsten carbide or synthetic sapphire nozzle, which is separated by a typical distance of 2 to 15 mm from the workpiece (see Fig. 10.1).

The mass rate of removal is usually about 10 mg min^{-1}; the air jet tool traverses the workpiece at a linear rate of 0 to 2 mm s^{-1}. The rate of material removal is affected by the air pressure, the size of the abrasive particles, the spray angle, the tool–workpiece separation distance, and the feed-rate.

Applications of AJM occur in the electronics industry for producing shallow and often intricate holes such as resistor paths in insulators and patterns in semi-conductors, and for engraving registration numbers on the toughened glass used for car windows.

10.3 CHEMICAL MACHINING (MILLING)

Chemical machining (CHM), also known as milling, is an etching process used to manufacture metal components to close accuracy.

An etchant-resistant mask, made typically of rubber or plastic, is used to protect those parts of the component from which no material is to be removed. Then the part to be machined is defined by any one of a number of procedures, such as scribing and removing the protective coating, exposure to light, and dissolving away the unexposed regions (De Barr, 1964), or by silk screen printing of the etchant-resistant layer.

Typical etchants are caustic and acid solutions, used for, respectively, aluminium and steel, and nickel and copper. Ferric chloride and hydrochloric acid are also used. For the chemical machining of copper, chromic acid and ammonium persulphate are used.

Figure 10.2 illustrates the main features of the technique. Undercutting

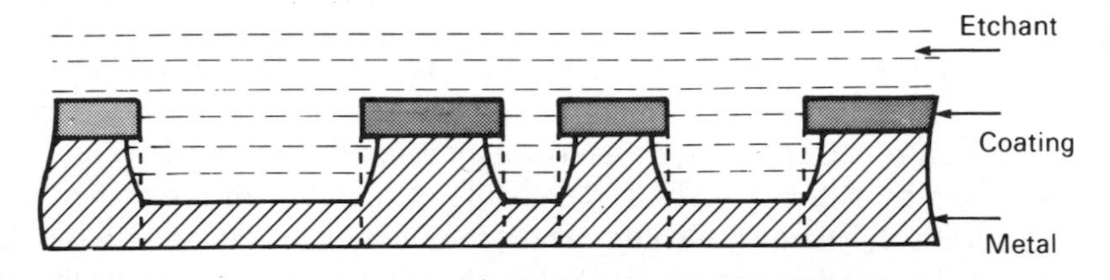

Fig. 10.2 Principles of chemical etching. (After De Barr and Oliver, 1968.)

occurs, the extent of which rises with the time that the component is immersed in the etchant. Metal removal rate is usually about 0.025 mm min^{-1} with surface finishes being about 30 to 125 μ inches Centre-Line Average (CLA) (0.75 to 3.75 μm). Dimensional tolerance is usually about the same magnitude as the thickness of metal removed.

CHM has a wide range of applications. It has been used for reducing the weight of aircraft wing panels, by removing metal from selected areas. Recently, CO_2 lasers have been employed for cutting the masks used in this technique in order to avoid the damage to the surface caused by manual cutting, see Fig. 10.3. The manufacture of printed circuits is another popular use of the technique, which is also applied for finishing to close tolerances and for the achievement of precision fits in items such as socket joints.

10.4 PHOTOCHEMICAL MACHINING (PCM)

Photochemical machining (PCM), sometimes known as photoetching, utilizes photographic and photoresist techniques coupled with wet chemical etching to remove material from specified areas of a workpiece substrate; it is especially attractive when the parts are complex and have to be produced in relatively thin sheet metal.

From Allen (1986), the process involves the following steps, summarized in Fig. 10.4. First, the required patterns are produced on a photographic film or glass plate, termed the phototool. Next, the sheet metal is chemically cleaned, and coated with a light-sensitive film of photoresist. The photoresist is normally negative-working; that is, it becomes insoluble in a developer, after it has been irradiated with actinic (ultraviolet) light. The photoresist is usually supplied in either liquid or dry film form. When the former type is used, the substrate has to be dipcoated, and then dried. If the latter is used, the dry film has to be laminated onto the sheet metal, by a combination of heat and pressure.

The next stage involves contact printing of the photo-tool image onto the photoresist, with ultraviolet light used as the actinic exposure source. The photoresist is then developed, enabling durable stencils to be formed on the metal.

By etching through the stencil apertures, usually with aqueous ferric or cupric chloride solutions to which have been added hydrochloric acid in carefully controlled amounts, the component is finally produced.

As shown by Allen, Fig. 10.5, the profile of the edge of the etched hole changes with the time of etching; when the etchant has dissolved the top surface of the metal and penetrates below the level of the stencil, etching can occur in any direction. Appreciable undercutting can arise through lateral

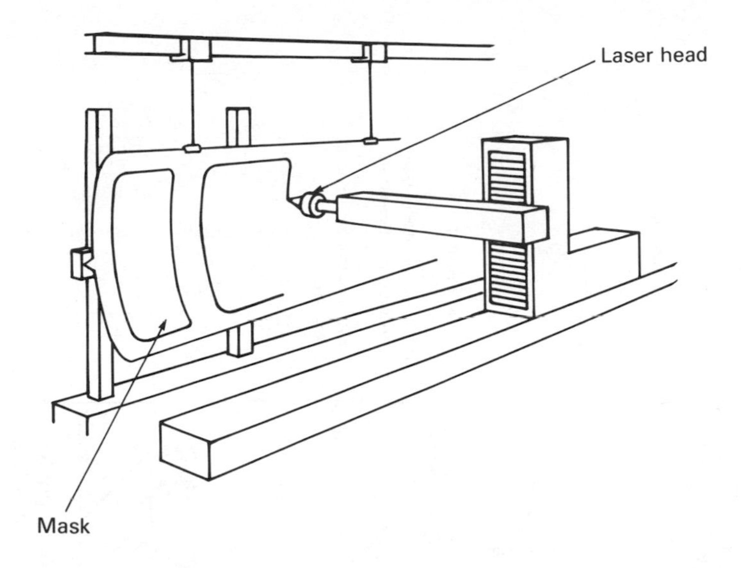

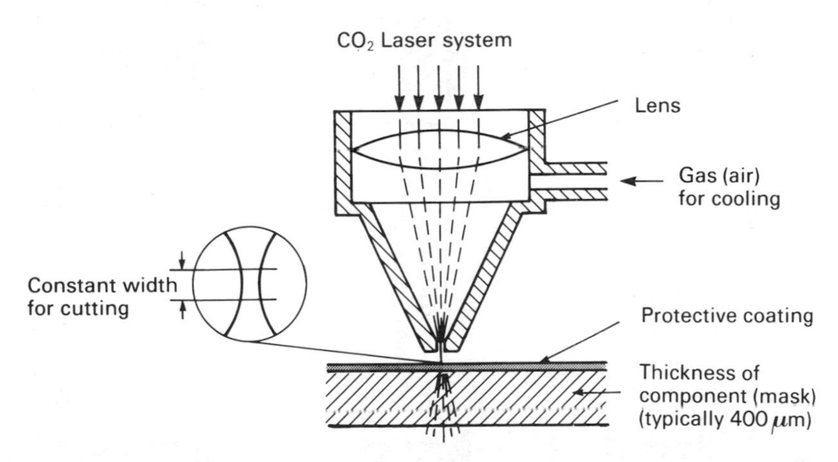

Fig. 10.3 Laser cutting of mask. (Laser power 75 kW.) (After Snoeys *et al.*, 1986.)

etching (Fig. 10.6). Convex, straight or concave profiles can be obtained by etching through mirror-image registered stencils. Tapered profiles, illustrated in Fig. 10.7, are possible by etching through dissimilar registered stencil apertures.

Since dimensions also vary with time of etching, the period of time needed to give the required profile has to be determined.

Allen reports that a typical tolerance is about ±0.05% of the dimension of

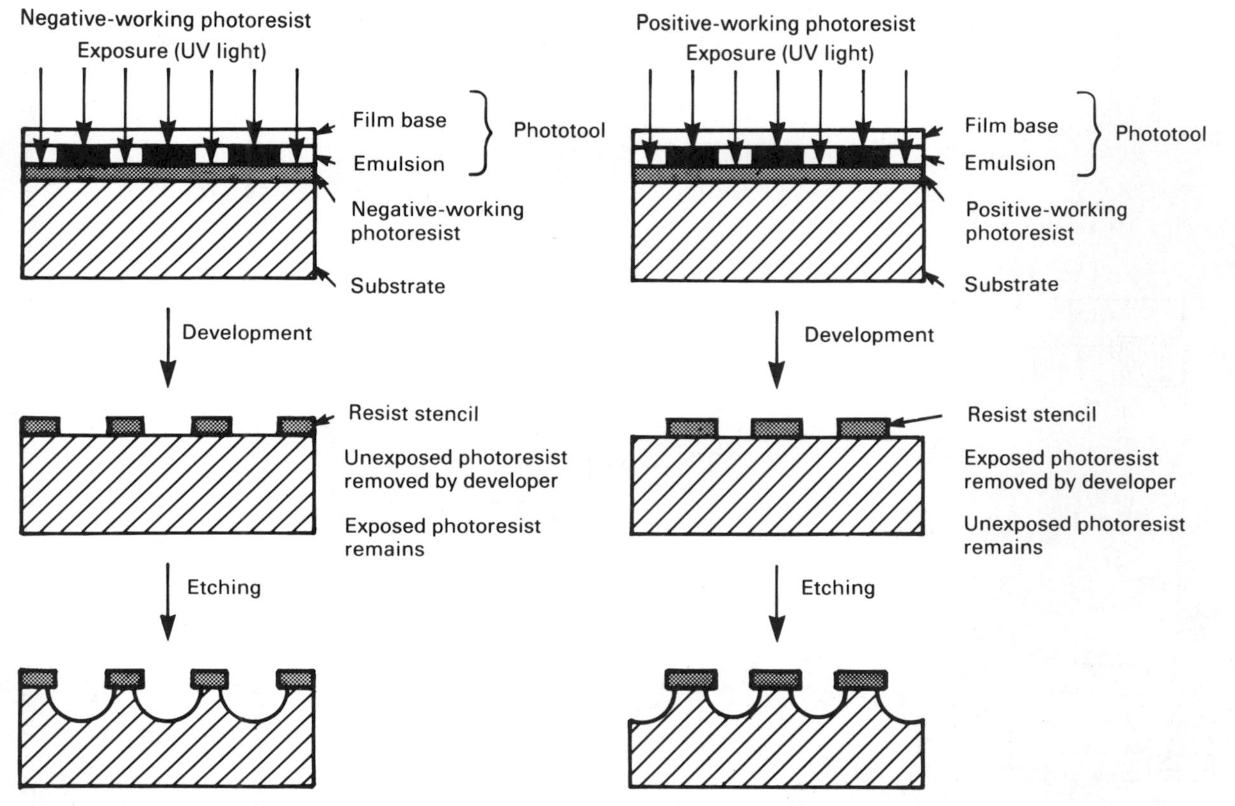

Fig. 10.4 Processing of substrates coated with negative- and positive-working photoresists. The phototool is of the same tonality in both cases. (By permission of Allen, 1986.)

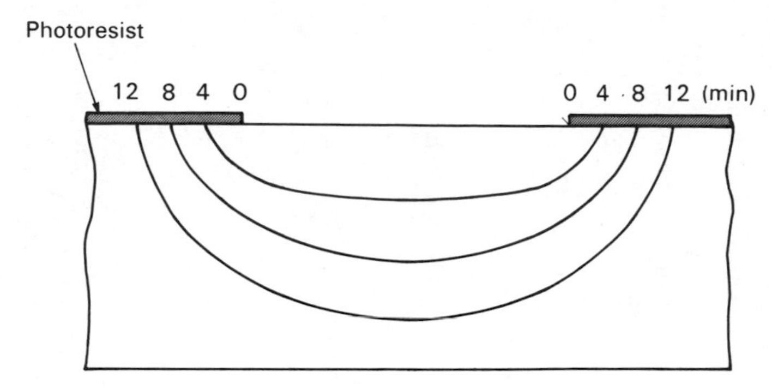

Fig. 10.5 Development of profile with etching time. (After Allen, 1986.)

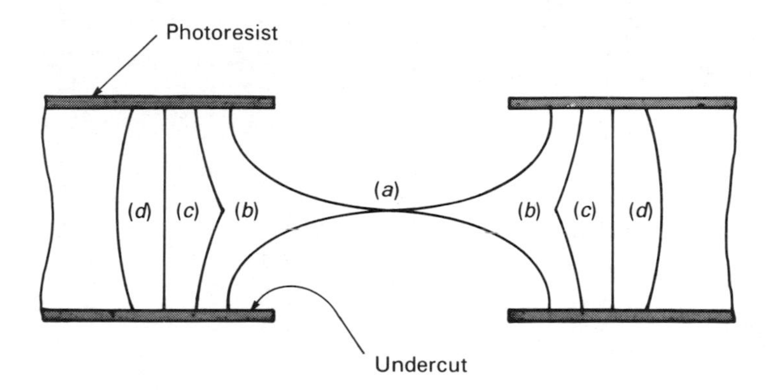

Fig. 10.6 Development of etched edge profiles. (After Allen, 1986.)
(a) Breakthrough point.
(b) Biconvex.
(c) Straight.
(d) Biconcave.

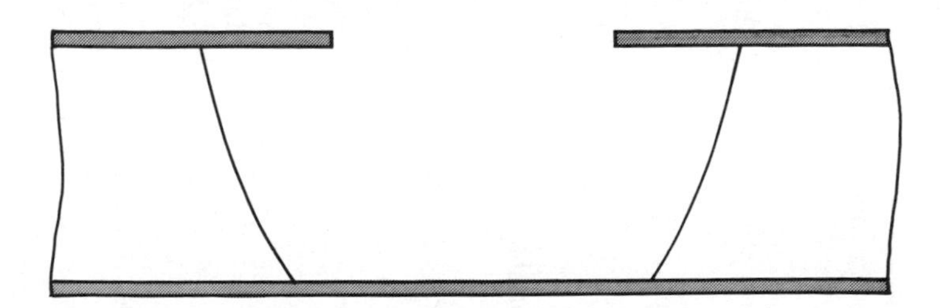

Fig. 10.7 Single-sided etch profile. (After Allen, 1986.)

Table 10.1 Technical comparison of fabrication techniques (Reproduced by kind permission of Allen, 1986)

	Photographic methods		*Machine shop methods*	
	PCM	*Photoforming*	*Stamping*	*Microdrilling*
Maximum material thickness	1.5 mm (6 mm for low resolution work)	2 mm	13 mm	Depends on drill size
Deviation from a straight profile	<20% of material thickness (T)	Dependent on material thickness (T) ro resist stencil thickness	A slight taper with a burr	A slight taper with a burr
Minimum aperture size	$\phi = 1.1T$ for most metals (but not an absolute limit)	$\phi = 0.1T$–$0.5T$	$\phi = 0.5T$ (low carbon steel) $\phi = 0.75T$ (high carbon steel)	$\phi = 0.025$ mm
Material	All metals (but vary in etchability)	Usually nickel, copper, silver or gold	Non-brittle metals	Non-brittle metals
Process advantages	(1) Produces burr-free and stress-free components (2) Physical and chemical characteristics of metal not altered during processing (3) Variable edge profile	May be the only viable process capable of achieving very high resolution	(1) Forming operations can be carried out whilst blanking (2) Fast	Fast
Process disadvantages	(1) A multi-stage process (2) Thickness limitation	(1) A multi-stage process (2) Restricted range of metals (3) Thickness limitation	(1) Long lead times (2) Deburring required	(1) Deburring required (2) Drill fragility (3) Difficult to locate drill (4) Skilled operatives required

the part, e.g. 50 ± 0.025 mm. Dimensional tolerances can usually be kept to about 10% of the metal thickness.

PCM can compete as a viable alternative with the most precise and fine stamping; components made by it are burr-free, and they retain the same chemical and physical properties as the original steel metal from which they are produced. The characteristics of PCM compared with stamping are given in Table 10.1.

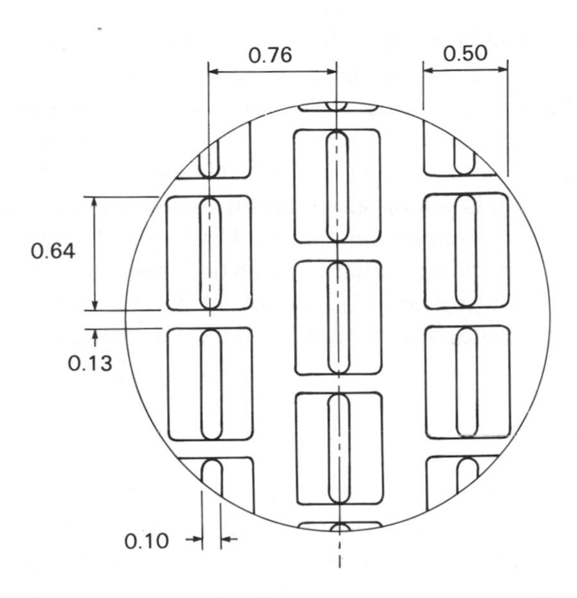

Fig. 10.8 Typical dimensions of a shadow mask (mm). (After Allen, 1986.)

The number of applications for PCM is rapidly widening. PCM has been used to etch 300 000 tapered slots in 0.15 mm thick mild steel, to be employed as a colour television receiver tube aperture mask (usually called the shadow mask). Typical dimensions produced are shown in Fig. 10.8. The process has also been chosen as a way to manufacture integrated circuit lead frames, on account of the ease with which phototool manufacture can replace hand tooling which shortens considerably the lead times.

Because stamping can affect adversely the magnetic permeability of materials, laminations to be used for recording heads are photo-chemically machined. Heat sinks, hybrid circuit pack lids and light copper disks are also made by this technique.

PCM is being increasingly used in the decorative and graphics industries for the production of signs and labels. Surface etching enables logos, instructions, part numbers and other alpha-numeric characters to be incorporated into designs. Etched fold lines enable a third dimension to be added to flat components for the fabrication of boxes and enclosures.

10.5 ELECTROGEL MACHINING

De Barr and Oliver (1968) describe the use of a semi-solid gel of cellulose acetate containing acid for machining surface cuts in honeycomb materials. The gel is first cut or cast to the complementary shape required in the workpiece, and is then placed on its surface. Metal is removed by the etching action of the acid; during machining the gel is allowed to advance into the honeycomb usually to a depth of about 10 mm. A low voltage can also be applied between the component and a metal-foil cathode placed on the surface of the gel. The consequential electrolytic action enables the ready disposal of the reaction products. The removal rate is increased to about 0.025 mm min^{-1}, a tolerance of about 0.025 to 0.075 mm being attainable.

10.6 ANISOTROPIC ETCHING

Snoeys, Staelens and Dekeyser (1986) have reported the rapid growth in the use of anisotropic etching techniques for the fabrication of micro-mechanical components. These techniques have been mainly applied to the manufacture associated with silicon integrated circuits. Their application to extremely small mechanical components which include electronic circuitry is a relatively recent development. The main difference is that the processing of silicon integrated circuits relies on a sequence of effectively planar processes, whereas for micromechanical components, material removal in three dimensions is needed in order to obtain a structure.

The technique may be explained firstly by consideration of the Miller indices of a crystal (integer multiples of the length of one edge of a unit cube). For crystalline silicon which has an interlocking face-centred cubic structure (Snoeys *et al.*, 1986) Fig. 10.9 shows the planes designated by the [110] and [100] Miller indices.

Some alkaline solutions at elevated temperatures when used as etchants are found to have etch rates that are much greater in certain crystallographic directions. In the case of silicon shown, the etchant makes a faceted hole in the crystal direction, perpendicular to the [110] plane, and less rapidly in the

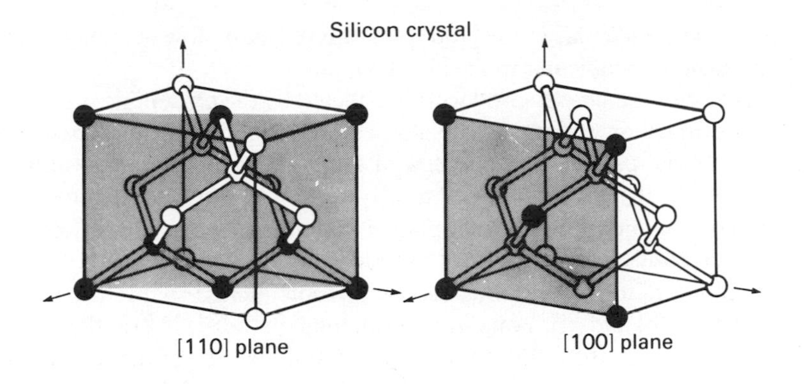

Fig. 10.9 Etching directions for crystalline silicon. (After Snoeys *et al.*, 1986.)

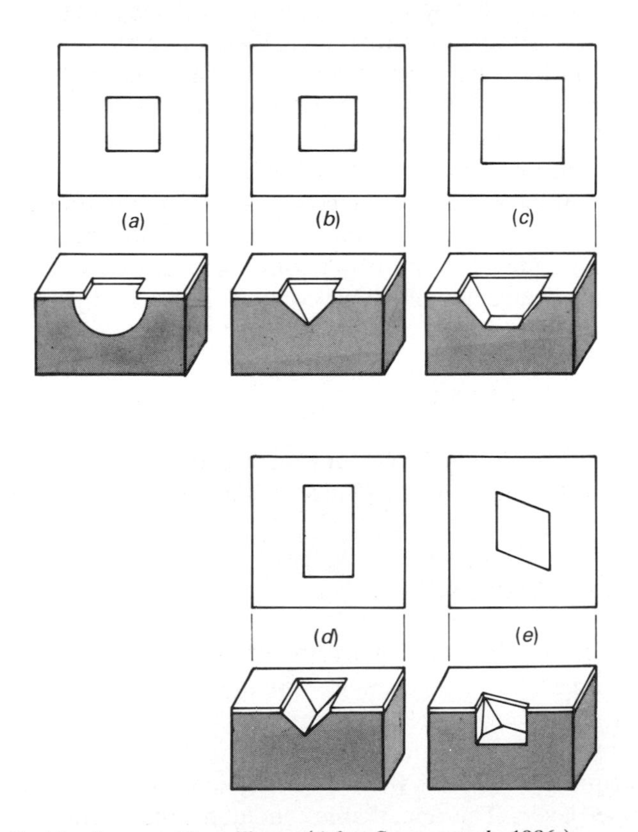

Fig. 10.10 Etching in crystalline silicon. (After Snoeys *et al.*, 1986.)

directions perpendicular to the [100] plane. Extremely slow etching occurs in the direction perpendicular to the [111] plane.

The shape of the hole so produced is affected by the crystalline orientation of the silicon wafer surface, by the shape and orientation of the openings in the mask at that part of the surface, and also by the particular orientation in which the anisotropic etchant is especially active. Anisotropic etchants cause faceted holes composed of crystal planes that are etched at the slowest rate; isotropic etchants produce a gently rounded hole (Fig. 10.10(a)). For a wafer with a plane of [100] Miller index some possible etched shapes are given in Fig. 10.10(b), (c) and (d). For a square opening on the mask orientated along the [110] direction of a [100] plane a pit in the shape of a pyramid with [111] side walls is produced. Figure 10.10(c) shows that a flat-bottomed pit is obtained when a larger mask opening is used, and when the etchant is halted before it reaches the [111] planes. Figure 10.10(e) shows the configuration obtained with a [110] orientation.

As an example of an application of anisotropic etching, Fig. 10.11 shows a piezoresistive accelerometer matrix composed of 16 silicon cantilevers of thickness 4 μm, formed by anisotropic etching. The cantilevers vibrate mechanically at resonant frequencies in the range of 4.3 to 6.4 kHz. Their mechanical deformation is converted into electric signals by means of polycrystalline piezoresistors. An attraction of arrays made in this way is the high sensitivity of the cantilevers which operate at their natural frequencies. They also offer a way of obtaining a Fourier-transformed frequency spectrum, without the need for further conversion. Anisotropic etchants are also used in the manufacture of optical components and ink jets.

10.7 ISOTROPIC ETCHING

Unlike the effects obtained with anisotropic etchants, isotropic etching produces a gently rounded hole, as illustrated in Fig. 10.10(a).

10.8 SELECTIVE ETCHING

Snoeys *et al.* (1986) discuss selectivity, an effect obtained with silicon highly doped with boron. The rate of etching of this material proceeds without being affected by the concentration of boron, up to about 2.5×10^{19} atom cm^{-3} when a 1000-fold reduction occurs in the etch-rate. This effect is used to obtain vertical structuring of silicon wafers. Thus high boron doped layers are used as etch-stopped layers. This remarkable condition is utilized in the fabrication

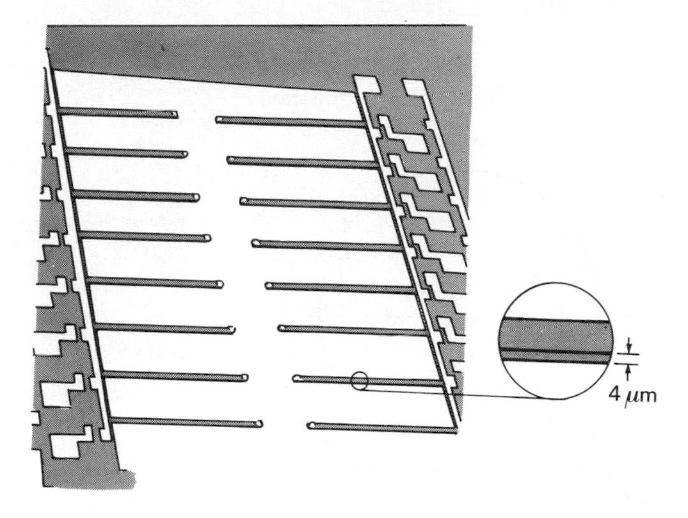

Fig. 10.11 Example of anisotropic etching of silicon cantilever. (After Snoeys *et al.*, 1986.)

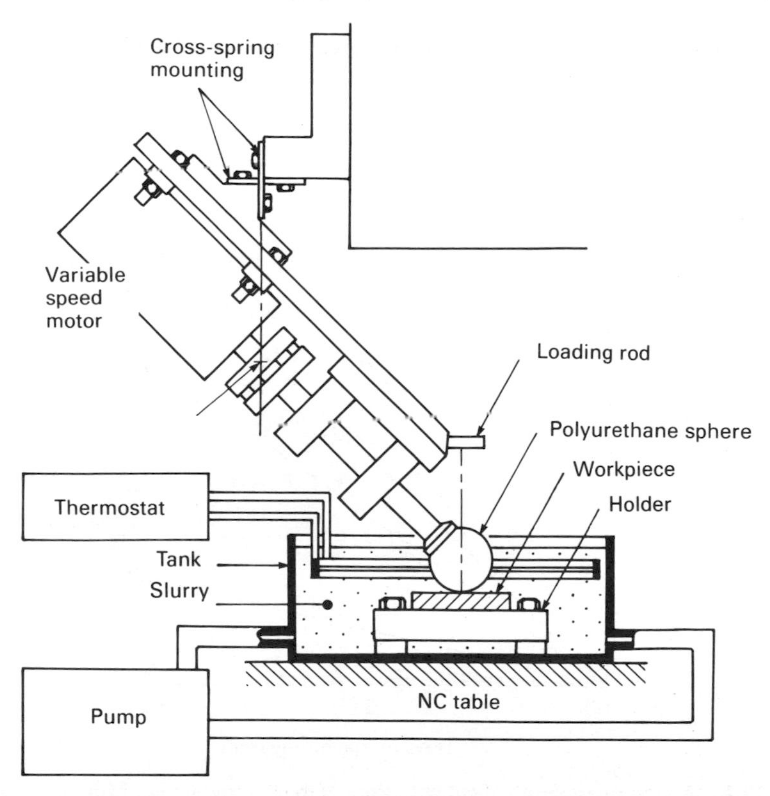

Fig. 10.12 Apparatus for elastic-emission machining. (After Mori *et al.*, 1987.)

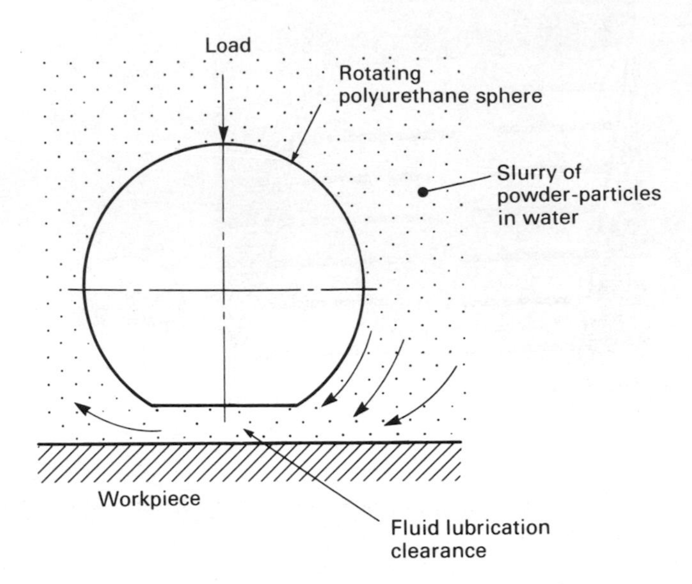

Fig. 10.13 Acceleration of powder slurry towards workpiece. (After Mori *et al.*, 1987.)

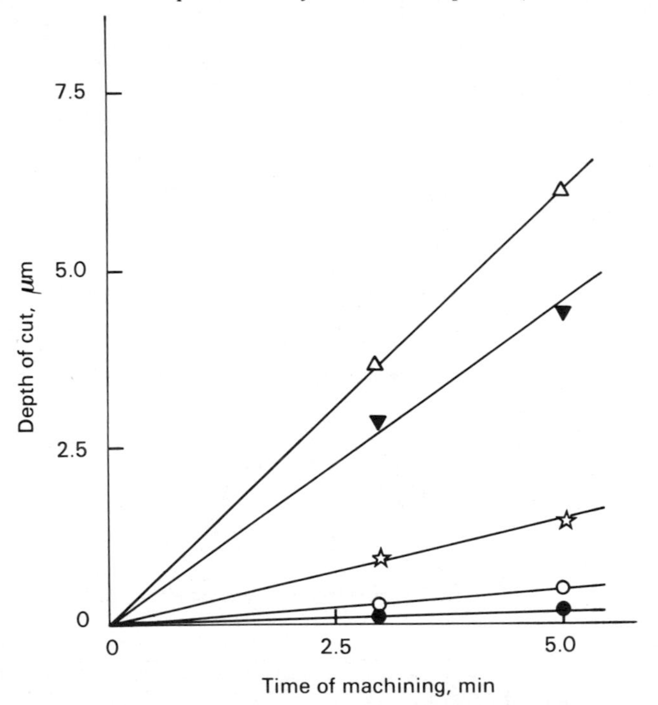

Fig. 10.14 Variation of depth of cut with time of machining. (After Mori *et al.*, 1987.)
△ Si(111), ▼ Si(100), ☆ Float glass, ○ Ge(100), ● GaAs(100).

of silicon membranes, and for springs and masks for X-ray, and electron- and ion-beam lithography.

Snoeys *et al.* (1986) describe a silicon membrane structured in this way, the membrane having spirals of thickness about 2.5 μm, with a depth of deflection of approximately 300 μm.

10.9 ELASTIC EMISSION MACHINING (EEM)

Mori, Yamauchi and Endo (1987) describe their investigations of this ultra-precise, atomic-size method of material machining, in which surface atoms are removed from a substrate by a process resembling chemical etching. This technique is based on the principle that when two solid materials are first placed together, and then separated, atoms of one surface may move onto the other, obtainable with the apparatus shown in Fig. 10.12. This condition of material removal is achieved by mixing ultra-fine powder particles, usually ZrO_2 of diameter much smaller than 1.0 μm, with water.

As indicated in Fig. 10.13, the particles are accelerated towards the work surface by means of a rotating polyurethane rubber sphere, cut to a tolerance of less than 1.0 μm, and of surface roughness of the order of 1.0 μm. The thickness of the fluid layer is typically 1.0 μm, which is much larger than that of the particles, and it is maintained so that the workpiece surface is not damaged.

Material is found to be removed from the workpiece over an area usually smaller than 10 nm^2. Figure 10.14 shows how the depth of cut rises with time of machining for a range of workpiece materials of interest to industry. Typical machining accuracy and surface finish are respectively better than 0.1 μm and 30 Å.

BIBLIOGRAPHY

Allen, D. M. (1986) *The Principles and Practice of Photochemical Machining and Photoetching*, Adam Hilger, Bristol.

De Barr, A. E. (1964) Article in *Machining and Manufacture*.

De Barr, A. E. (1966) Modern Metal Removing Techniques, *The Chartered Mechanical Engineer*, March, No. 3, 106–13.

De Barr, A. E. and Oliver, D. A. (1968) (Eds) *Electrochemical Machining*, MacDonald Press, UK, chp. 2.

Gunter, D. (1984) Photochemical Machining, *Engineering (London)*, **224** (2), I–IV.

Mori, Y., Yamauchi, K. and Endo, K. (1987) *Elastic Emission Machining*, Paper presented at 3rd Int. Prec. Eng. Seminar, Cranfield, UK.

Snoeys, R., Staelens, F., Dekeyser, W. (1986) Current Trends in Non-conventional Material Removal Processes, *Ann. CIRP*, **35** (2), 467.

Taniguchi, N. (1983) Current Status in, and Future Trends of, Ultraprecision Machining and Ultrafine Materials Processing, *Ann. CIRP*, **32** (2), 1–8.

Appendix: Basic atomic and electrical principles

A.1 INTRODUCTION

A useful foundation for understanding many advanced methods of machining can be obtained from an appreciation of electrical phenomena together with the bases of atomic and electronic properties of matter. This appendix therefore is devoted to such a presentation.

A.2 BASIC PROPERTIES OF ELECTRICITY

The word 'electricity' itself is obtained from the Greek for 'amber'. Thales of Miletus (640 to 548 BC) noted that when he rubbed a piece of amber, particles of matter were attracted to it. The study of the properties of electricity at rest, or electrostatics as it is called, has stemmed from this observation.

Electricity itself may be regarded as being composed of two kinds of charge: positive and negative. The smallest negative charge is the 'electron', discovered by J. J. Thomson in 1897. The electron has a mass of 9.0×10^{-31} kg, and a radius of approximately 1.9×10^{-15} m. Its charge is 4.80×10^{-10} electrostatic units (esu) or 1.6×10^{20} electromagnetic units of charge.

A negative charge is attracted to a positive one; like charges are found to repel each other. The smallest positive charge, which has the same numerical value as that of the electron, is found on a particle called the proton. However, the mass of the proton is about 1837 times heavier than that of the electron.

The radius of an electron is about (1/50 000)th of that of the atom, the basic particle of which matter is composed. The atoms that make up matter do not normally carry any excess amounts of positive or negative charge; that is, the atoms are neutral. About 10^{23} atoms are present in 1 ml of an ordinary solid. The atom itself consists of positive and negative charges associated together in small structures of about 10^{-8} cm in diameter. The positive charge occurs on the 'nucleus' of the atom, where most of its mass is also concentrated. The nucleus in turn is composed of protons, each with a charge equal to that of an electron, but of opposite sign, and neutrons which are uncharged. The number of protons is usually equal to, or greater than, the number of neutrons. Since the mass of both the proton and neutron is about 2 000 times greater than that of the electron, the negatively charged particles in the atom, in the form of electrons, may be regarded as moving around the nucleus in a planetary-like orbit, a concept first proposed by Rutherford in 1911. (The exact manner of these movements such as wavelike motion instead of an orbiting system is not discussed here.)

Based on these observations, Bohr deduced an expression for the energy of an electron moving around the nucleus.

A.3 BOHR'S THEORY OF ATOMIC STRUCTURE AND RADIATION

First consider an atom to consist of a nucleus of charge $+Ze$ and mass M and a single electron of charge $-e$ and mass m (for example, for a hydrogen atom $Z = 1$, for a singly ionized helium atom, $Z = 2$ etc.). The electron is assumed to rotate in a circular orbit about the nucleus. The mass of the electron is assumed to be negligible in comparison with the mass of the nucleus; thus the nucleus can be assumed to remain fixed in space (Fig. A.1).

Bohr assumed that the Coulombic force acting on the electron is balanced by the centrifugal force needed to keep it in its circular orbit. That is

$$\frac{Ze^2}{4\pi\epsilon_0 r^2} = \frac{mv^2}{r} \tag{A.1}$$

where v is the velocity of the electron in its orbit, and ϵ_0 is the permittivity of free space, r being the radius of the orbit.

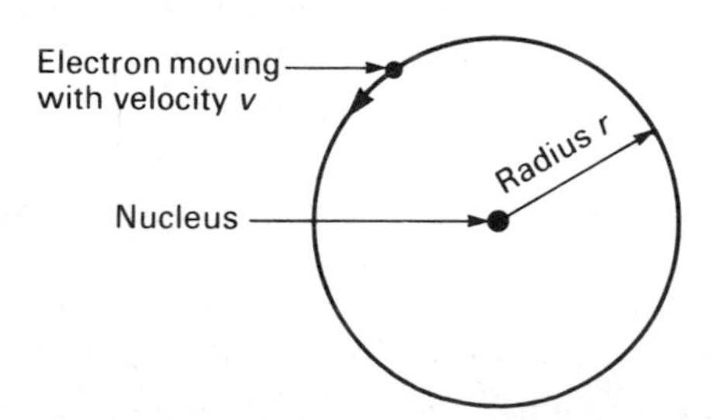

Fig. A.1 An electron moving in circular orbit around nucleus.

Next the orbital angular momentum of the electron, mvr, was, Bohr postulated, only able to have certain values given by an integral multiple of Planck's constant divided by 2 (instead of having an infinity of orbits which would be possible in classical mechanics). That is,

$$mvr = \frac{nh}{2\pi}, n = 1, 2, 3 \tag{A.2}$$

On substitution in equation (A.1) we find that

$$r = \frac{n^2 h^2 \epsilon_0}{\pi m Z e^2} \quad n = 1, 2, 3 \tag{A.3}$$

and so

$$v = \frac{Ze^2}{2nh\epsilon_0} \tag{A.4}$$

Next, the energy of the electron consists of kinetic energy, $1/2\, mv^2$, and potential energy. The latter quantity can be evaluated from work needed to move the electron from its position r to that position ($r = \infty$) at which the potential energy is zero. Thus the potential energy is

$$\frac{-Ze^2}{4\pi\epsilon_0 r} = -mv^2 \tag{A.5}$$

The total energy at position r, relative to that at infinity, when the electron is at rest is then $-1/2\, mv^2$.

Since

$$v = \frac{Ze^2}{2nh\epsilon_0} \tag{A.6}$$

the total energy is

$$-\frac{1}{2}mv^2 = \frac{-Z^2 me^4}{8n^2 h^2 \epsilon_0^2}$$

$$= -2.18\frac{Z^2}{n^2} \times 10^{-18}\ \text{J} \tag{A.7}$$

(On insertion of the values $m = 9.1 \times 10^{-31}$ kg, $e = 1.6 \times 10^{-19}$ C, $h = 6.625 \times 10^{-34}$ J s, $\epsilon_0 = 8.854 \times 10^{-12}$ F m^{-1}.)

The quantity n is known as the 'principal quantum number'. The state of lowest energy is given by $n = 1$, and is known as the 'ground state'. Higher energy levels, for which $n = 2, 3 \ldots$ etc., are said to be 'excited states'.

When an electron moves from an outer orbit of energy E_1, to an inner one with energy E_2, the decrease in its total energy, δE, is equal to the difference in its energies in the two orbits:

$$\delta E = E_1 - E_2 = 2.18 \times 10^{-18}\, Z^2 \left(\frac{1}{n_2^2} - \frac{1}{n_1^2} \right) \mathrm{J} \tag{A.8}$$

This energy is emitted as a particle or quantum of electromagnetic radiation of frequency ν where

$$\nu = \delta E / h \tag{A.9}$$

or wavelength λ where

$$\lambda = \frac{c}{\nu} = ch/\delta E \tag{A.10}$$

A.4 OTHER QUANTUM NUMBERS

Bohr's theory introduced the principal quantum number n as a measure of the energy of the electron. A full description of the state of the electron requires the introduction of other quantum numbers.

The angular momentum quantum number, l, allows for the possibility of elliptical, in addition to circular, orbits. A restriction is again placed on the possible orbits. Thus

$$l = 0, 1, 2, \ldots (n-1) \tag{A.11}$$

The magnetic quantum number, m_l, takes account of restriction of possible directions of the axes of the electron orbits:

$$m_l = -l, -(l-1), \ldots, -1, 0, 1, \ldots l \tag{A.12}$$

The spin quantum number m_s caters for possible values of the spin on the electron

$$m_s = \pm 1/2 \tag{A.13}$$

Example

1. If $n = 1$, then $l = 0$, $m_s = \pm 1/2$. Thus there are only two sets of quantum for an electron at this energy level.
2. If $n = 2$, then, either $l = 0$, $m_e = 0$, $m_s = \pm 1/2$;
   or $l = 1$, $m_e = -1, 0, +1$, $m_s = \pm 1/2$.

That is eight sets of quantum numbers are available. In general, for any value of n, there are $2n^2$ possible sets of quantum numbers.

Based on these results Fig. A.2 indicates the wavelengths and energy levels associated with quantum numbers for the hydrogen atom.

This understanding of the energy associated with the electron movement around the nucleus will help when laser machining is considered.

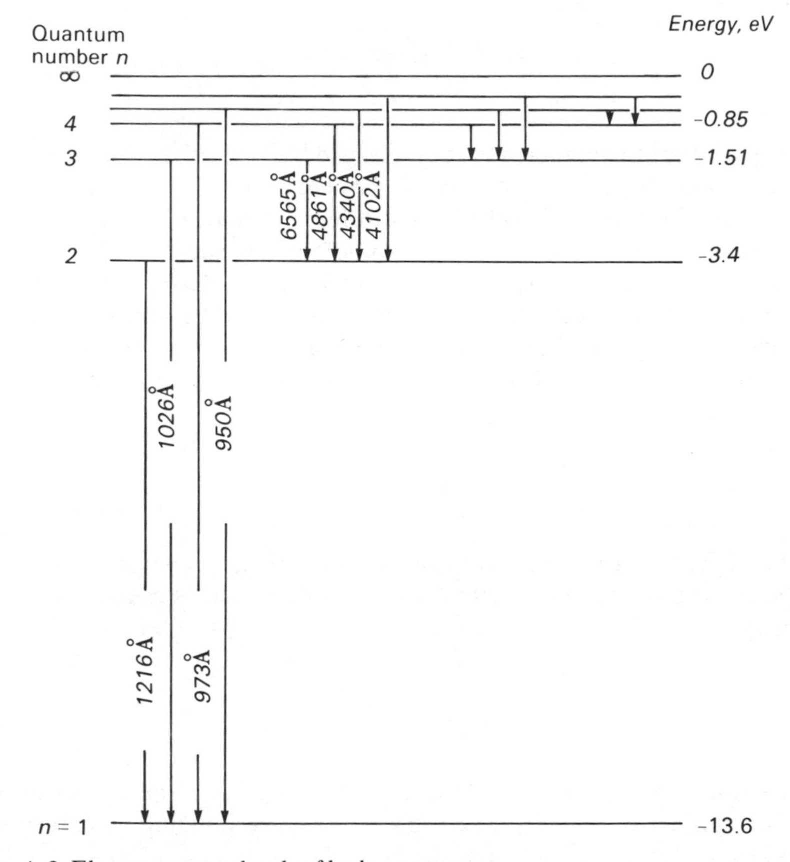

Fig. A.2 Electron energy levels of hydrogen atom.
Note: 1 eV (electron volt) = 1.6×10^{-19} J
1 Å (Ångström) = 10^{-10} m.

A.5 DUAL NATURE OF MATTER

A review of other aspects of the basic properties of matter, especially its dual nature of waves and particles, is useful to a study of advanced methods of machining.

Einstein supported the view that light and other types of electromagnetic radiation are comprised of pulses of electromagnetic waves. These pulses are termed 'photons'. They are emitted or absorbed as discrete entities, each having an energy $h\nu$, where ν is the frequency of the electromagnetic waves. Figure A.3 shows the relationship between frequency and photon energy.

In 1924, De Broglie postulated that matter in general might be subject to wave–particle duality, showing that the wavelength λ associated with a particle of matter would be represented by

$$\lambda = \frac{h}{mv} \tag{A.14}$$

where mv is the momentum of the particle, and h is Planck's constant. That is, the momentum of a photon and a particle would be h/λ.

Based on this hypothesis, other work, including that by Einstein, Davisson and Germer and Č. P. Thomson, led to proof that an electron beam can also be diffracted by crystals, just like X-rays. Indeed, from the crystal structures of metals which were already determined from X-ray diffraction studies, the wavelengths of electron beam motion were deduced.

The energy $1/2\ mv^2$ of the electrons in a beam was deduced to be given by the expression

$$\frac{1}{2}mv^2 = eV \tag{A.15}$$

where e is the charge on an electron, and V is the voltage used to accelerate it.

The corresponding wavelength λ based on de Broglie's postulate is then given by

$$\lambda = \frac{h}{mv} = \frac{h}{\sqrt{meV}} \tag{A.16}$$

For example, on substitution of typical values for h, m and e, and for an electron subjected to a voltage of 15 kV, the corresponding wavelength is calculated to be 10^{-11} m.

This behaviour applies also to other particles, such as protons, neutrons and ions.

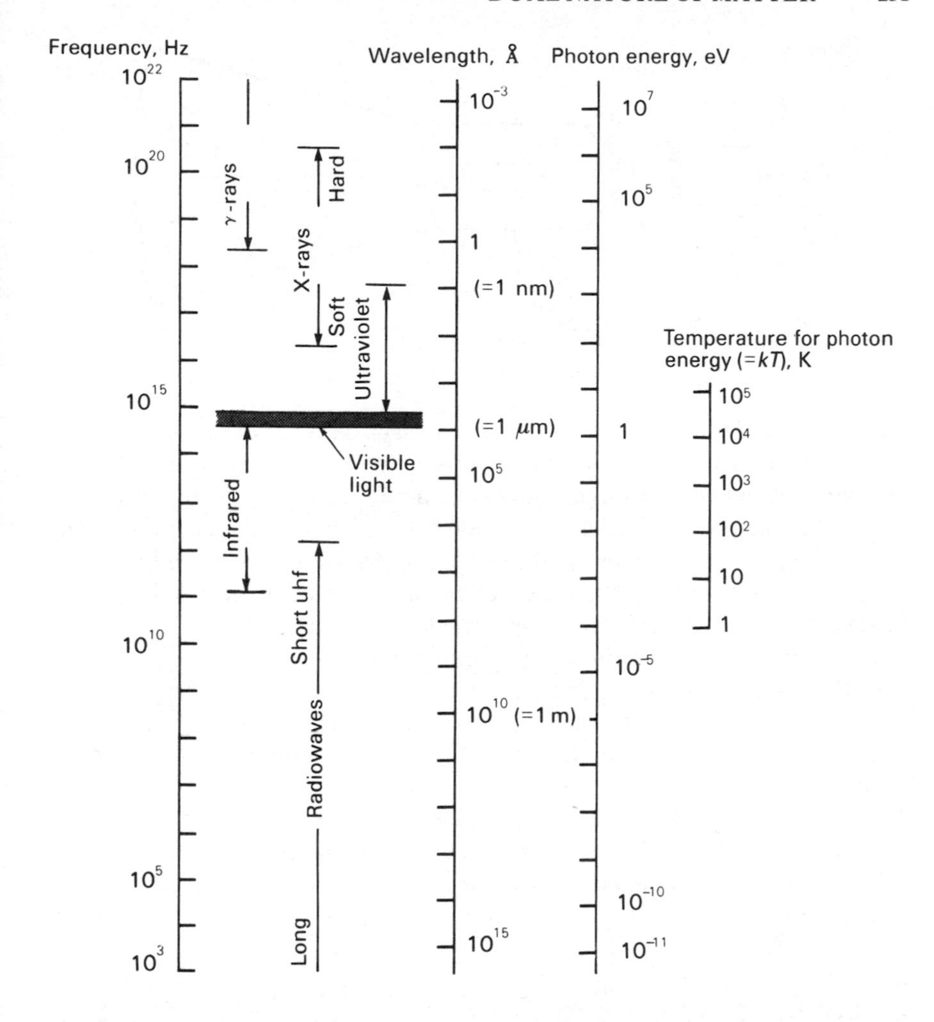

Fig. A.3 Relationship between frequency and photon energy.

Further relevant information for advanced machining can be obtained from experiments designed to measure the ratio of the charge to the mass of an electron, since some of these machining techniques work on similar principles.

J. J. Thomson used a discharge tube, such as that shown in Fig. A.4. A high voltage is applied between a cathode and a perforated anode, electrons leaving the former electrode and penetrating the latter. The electrons then travel through a hole in a screen B. This screen narrows the beam width to a shape resembling a thin pencil. The beam next traverses between metal

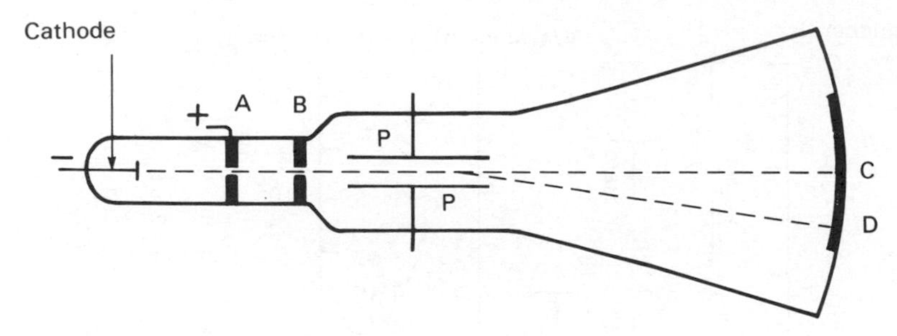

Fig. A.4 Discharge tube based on Thomson's work.

plates, denoted by PP in Fig. A.4, across which an electric field can be applied. The electron beam simultaneously passes between the poles of an electromagnet, not shown in the figure, but considered to be positioned above and below the plane of the paper. If neither field is applied, the electron beam strikes the point C in the centre of a screen, composed of fluorescent material at the end of the tube. If either the electric or electromagnetic field is applied the beam pencil is deflected to some other location, indicated by D.

Electron and ion beam machining equipment work on these principles, with the appropriate particles accelerated from a cathode towards a target (the workpiece) by means of a high applied voltage. Material is removed either by vaporization, in the case of electron beam machining, or by dislodgement of actual ions, by the ion method.

With so many advanced methods of machining relying on electrical effects for their function, further consideration of these matters becomes appropriate. Conductors of electricity, and insulators, are therefore next considered.

A.6 CONDUCTORS

Whereas electrostatics is concerned with electric charges at rest, the study of electric currents deals with charges in motion. The current may arise from motion of either positive or negative charges, or both. For example, in metals, the charge is transported by electrons.

On a microscopic scale, these charges move in all directions at different speeds; macroscopically, the average motion of the charges within a very small volume is used to determine a mean drift velocity, from which the magnitude and direction of the electric current are obtained.

In the discussion, materials may be regarded as either conductors or insulators. In the former, the electric charge can flow readily, since the electrons are easily separated from their associated positive charges, and are therefore free to move about the whole volume of the material, usually under the influence of a force, such as an applied voltage. All metals exhibit this property, as do some other substances such as graphite, which is used as the tool-electrode in electro-discharge machining.

Thus in metals, of which copper is a common example, a certain number of electrons, called 'metallic', or 'free', or 'conduction' electrons, are able to flow freely through the material, giving rise to current; the positive charges remain fixed.

Another distinct class of conductors are those termed 'electrolytes', of which common salt (NaCl) in water is an example pertinent to the chapter on electrochemical machining; alternatively, the dilute sulphuric acid used in an ordinary accumulator is another common type of electrolyte. Each particle, or molecule, of the electrolyte spontaneously separates into positive and negative parts, which move independently of each other. Thus for NaCl solution, the sodium ions (that is, charged atoms) move in one direction carrying a positive charge; the chloride ions travel in the opposite direction, bearing a negative charge. The total current is the sum of two separate currents.

A.7 CURRENT FLOW

For a steady current, that is, one flowing independently of time, suppose that the charges move with mean velocity v_m. If there are N_c charged particles per unit volume, each carrying charge e, the current density J is defined by

$$J = N_c e v_m \qquad \text{(A.17)}$$

The direction of J is that in which the current flows.

A typical velocity v_m of the charges is of the order of 10^{-2} ms^{-1} (for a current of 1 A passing along a wire of 1 mm^2 section); cf. the random velocities of the electrons, which are of the order of 10^5 ms^{-1}.

That is, even a large current flow hardly affects the general distribution of velocities; thus quantities such as the time between collisions are not affected by current. Also the mean velocity v_m of the charges lies in the same direction as the field E that causes the motion. That is the current density J and the electric field E are parallel vectors.

Moreover experiments show that J is proportional to E, which is Ohm's law:

$$J = \kappa E \qquad \text{(A.18)}$$

where κ is the conductivity of the metal, the magnitude of which is not affected by the shape of the conductor, although it is dependent on temperature. The quantity $1/\kappa$ is called the specific resistance of the material.

Equation (A.18) is sometimes also expressed in the following, more familar way. If a current I enters one electrode of a conductor and leaves at another, and if V is the potential difference between the two electrodes, the ratio

$$R = \frac{V}{I} \tag{A.19}$$

is known as the resistance of the conductor. For example, for a straight wire of length l and constant cross-section A, and plane electrodes normal to the wire with potentials V_1 and V_2, the current density J is equal to

$$\frac{\kappa(V_1 - V_2)}{l}$$

The total current

$$I = JA = \frac{\kappa A(V_1 - V_2)}{l}$$

Thus from equation (A.19) the resistance is $1/\kappa A$. This concept of resistance is useful in understanding the bases of electrochemical machining.

A.8 INSULATORS

Practically no current can flow in an insulator, as all the negative charges (the electrons) are firmly attached to their corresponding positive charges. Glass, light oil, paraffin are examples of insulators. The last two mentioned feature in electrodischarge machining, as the dielectric fluid.

Faraday discovered that if the space between the two plates of a condenser was filled with a dielectric such as glass, its capacity was multiplied by a constant, the magnitude of which depended on the material, but which was not affected by the shape of the condenser. The relative values of this dielectric constant, as it is termed, range from one for free space, to six for glass, to about eight for water.

When an electric field is applied to a dielectric the positive and negative charges on each atom of the material are pulled in opposite directions. If the atom has no resultant charge on it, there should be no net force, and the effect of the applied field is the slight separation of the positive from the negative charges. The dielectric is said to be polarized by the electric field E. Each atom or molecule in the dielectric becomes a tiny dipole, the strength of which is proportional to the magnitude of the electric field (if the latter is not too large). The constant of proportionality is termed the polarizability, say α. The quantity αE is called the induced dipole. Another contribution to the polarizability arises when the molecules of the dielectric are permanent dipoles. Water is such a case. When a field is applied a couple is developed which tends to orient the water molecules relative to the field. A moment in the direction of the field occurs. If this effect has been caused by polarizability, and if there are N molecules per unit volume, the total moment per unit volume induced by the electric field is given by $N\alpha E$. This vector quantity is called the polarization, P. That is

$$P = N\alpha E = kE \tag{A.21}$$

where the constant k ($=N\alpha$) is called the dielectric susceptibility, a quantity noted to be dependent on the density of the material (given by N) and on the polarizability.

A.9 CONVERSION OF ELECTRICAL ENERGY TO HEAT

When a current flows, the electrons fall from regions of higher to lower potential. That is, electrical energy is lost. This loss manifests itself in the generation of heat.

For an electrode at potential V from which the rate of flow of electrical charge per unit time is I, the electrical power provided is VI. If the current is assumed to be steady, this must also be the rate at which the electrical energy is converted to heat.

The means by which this change in energy arises can be understood from the collision mechanisms of a conduction electron in a metal. Now the average time between its successive collisions and the non-conducting material of the metal is approximately 2×10^{-13} s. During this time, each conduction electron acquires a forward momentum in the direction of the current. On collision, this momentum is transferred to the fixed material of the metal, the nuclei of which will therefore vibrate with more energy. From kinetic theory, this energy of vibration can be described as heat, exhibited as a rise in temperature.

Many advanced methods of machining, including the laser, electron beam and plasma arc techniques, rely on the production of heat by transfer of energy in order to achieve material removal.

BIBLIOGRAPHY

Coulson, C. A. (1951) *Electricity*, 2nd edn, Oliver and Boyd, Edinburgh.

Crowther, J. A. (1957) Ions, Electrons and Ionising Radiations, 8th edn, Edward Arnold, London.

Flinn, R. A. and Trojan, P. K. (1986) *Engineering Materials and Their Applications*, 3rd edn, Houghton Mifflin, Boston.

Pascoe, K. J. (1972) *An Introduction to the Properties of Engineering Materials*, 2nd edn, Van Nostrand Reinhold, London.

Pascoe, K. J. (1973) *Properties of Materials for Electrical Engineers*, John Wiley & Sons, London.

Author index

Page numbers in **bold** refer to tables, in *italics* to figures.

Subject index

Page numbers in **bold** refer to tables, in *italics* to figures.